D1002180

A Wonderful
Life

A

W O N D

HARPER
DESIGN
An Imprint of HarperCollins Publishers

INSIGHTS ON FINDING A
MEANINGFUL EXISTENCE

ERFUL

FE

FRANK MARTELA

A WONDERFUL LIFE:
Insights on Finding a Meaningful Existence
Copyright © 2020 by Frank Martela.

All rights reserved. No part of this book may be used or reproduced in any manner whatsoever without written permission except in the case of brief quotations embodied in critical articles and reviews. For information address Harper Design, 195 Broadway, New York, NY 10007.

HarperCollins books may be purchased for educational, business, or sales promotional use. For information, please email the Special Markets Department at SPsales@harpercollins.com.

First published in 2020 by
Harper Design
An Imprint of
HarperCollins*Publishers*
195 Broadway
New York, NY 10007
Tel: (212) 207-7000
Fax: (855) 746-6023
harperdesign@harpercollins.com
www.hc.com

Distributed throughout the world by
HarperCollins*Publishers*
195 Broadway
New York, NY 10007

ISBN 978-0-06-294277-7
Library of Congress Control Number:
2019027330

Book design by
Roberto de Vicq de Cumptich

Printed in Canada
First Printing, 2020

"HOW

"How did I get into the world? Why was I not asked about it, why was I not informed of the rules and

into the

regulations but just thrust into the ranks . . . ? How did I get involved in this big enterprise called actuality?

Why should I be involved? Isn't it a matter of choice? And if I am compelled to be involved, where is the

did I get

manager—I have something to say about this. Is there no manager? To whom shall I make my complaint?"

—SØREN KIERKEGAARD,
Repetition, 1843

world?"

here were you when the meaninglessness of life hit you? Was it over your third microwaved dinner of the week while pondering the flavor and health benefits of ketchup? What about when you hit the send button at 2 A.M. having completed an urgent work task only to realize that the world, most likely, won't be an inch improved in light of your accomplishment? Maybe a life-changing tragedy made you realize that you haven't put in the effort to contemplate what you really want out of life. Or perhaps you simply woke up one morning, stared at yourself in the bathroom mirror, and wondered if there was anything more to this crazy little thing called life.

Don't worry—you're not alone. In this book, you'll be in the company of many great thinkers and philosophers who came face-to- face with the insignificance of existence and ended up on the other side with a life-affirming, revitalized sense of meaningfulness.

As humans, we yearn for our lives to matter, be valuable, and have meaning. We're "hardwired to seek meaning," as professor of psychology Roy Baumeister has argued.[1] Lack of meaning is a serious psychological deprivation associated with depression and even suicide.[2] Meaning is important for human motivation, well-being, and, more generally, to having a life considered worth living.[3] Indeed, several studies have shown that those individuals who experience a strong sense of purpose in life tend to live longer.[4] Viktor Frankl, Holocaust survivor and noted psychiatrist, observed this firsthand during his time in the concentration camps. Only those individuals who were able to retain a sense of purpose in such intolerable conditions had a chance of surviving. He was fond of quoting Nietzsche: "He who has a why to live for can bear almost any how."[5]

The trouble is that Western culture has become increasingly incapable of answering the inevitable "why" question in any real or satisfying way. Throughout history, most civilizations have answered the yearning for meaning by providing a stable cultural framework that includes answers to life's biggest questions. When our ancestors asked, "How should I live my life?", they turned to their culture—society's staid-and-stable customs, beliefs, and institutions—to provide guidance. Today's cultural era, however, has destabilized the old foundations of meaning. While modern science has vastly improved the material conditions of life, it's taken the steam out of old-world value systems and explanations and failed to provide a new solid foundation for human values and meaning. As the expert on the history of morality, Scottish philosopher Alasdair MacIntyre, argues, modern Western values are built on the fragments of an older worldview that no longer makes sense.[6] Western societies have inherited certain values but have lost contact with the broader worldview that used to ground and justify them. And the influence of this increasingly secular and individualistic Western worldview is currently growing stronger around the world.[7]

In the idealistic modern worldview, you are free to source your own sense of meaning, and blaze your own unique path, based on your self-selected values. Unfortunately instead of feeling liberated, you just feel empty. You work harder, smarter, and more effectively than previous generations, but you're increasingly at a loss in explaining why you push so hard. To what end does your tedious labor serve? You've willingly fallen into the "Busy Trap" that author Tim Kreiner so eloquently describes: "Busyness serves as a kind of existential reassurance, a hedge against emptiness; obviously your life cannot possibly be silly or trivial or meaningless if you are so busy, completely booked, in demand every hour of the day."[8] You do whatever you can to retain this sense of busyness and urgency to avoid boredom and the threat of being alone with your thoughts. People seem willing to pursue whatever goals authority figures prescribe them in order to avoid thinking about what they themselves truly want to do with their lives. This explains the oddity about modern existence that philosopher Iddo Landau noted: "Many dedicate more thought in one evening to deliberating which restaurant or film they should go to than they do in their entire lifetime to deliberating what would make their lives more meaningful."[9]

To live a self-chosen life, to steer your own ship, you need to have a clear idea as to what direction you want to take. For that, you need some core values to help navigate life's challenges. And *for that,* you need to take some time to contemplate and question your life choices and face any existential doubt that may linger beneath the surface of your existence. There's a long and storied history of thinkers—from Leo Tolstoy and Thomas Carlyle to Simone de Beauvoir and Søren Kierkegaard to Alan W. Watts—who found that it's only by facing the absurdity of life and embracing the insignificance of existence that you can liberate yourself to find a more solid sense of meaning in your life. This book offers a new way to think about meaningfulness that speaks to our common humanity so that, no matter where you come from culturally, religiously, or otherwise, you will be guided toward a more fulfilling, meaningful life.

I want to help you live a more meaningful existence. After a decade of research into the philosophy, psychology, and history of the meaning of life, I've realized that identifying what makes life meaningful is easier than you think. In fact, there's probably an abundance of meaningfulness in your life, if you open yourself to seeing it and feeling it. The reason why the meaning of life often feels like an impossible and agonizing riddle is that we, as a culture, have continued to use old models to think about the question that no longer makes sense. By changing the way you think, you'll come to see that the answers you seek can be sourced from your everyday life. This book explains why humans seek meaning in the first place, examines the historical mistake that has given rise to modern existential anxiety, and offers you easily applicable paths that lead towards a more meaningful existence. Some of the insights might seem strange, others may be obvious, and still others you may already fully embrace. Together, however, they aim to offer a solid and stable foundation for you to build a fulfilling, life-affirming, and more meaningful existence.

YOU DIDN'T CHOOSE TO BE BORN.

You didn't choose to be born. No one asked your permission to be involved. No one provided you with an instruction manual and yet here you are, thrown into the world in which you need to act, to make something meaningful out of the limited time of existence afforded to you.[10] And you'd better figure it out soon, before it's too late. As Edward Norton's character, the narrator, says in the movie *Fight Club*, "This is your life, and it's ending one minute at a time."

Part

WHY

Humans

SEEK One: Meaning

Rise Above Life's

ABSU

"It happens that the stage sets collapse. Rising, streetcar, four hours in the office or the factory, meal, sleep, and Monday Tuesday Wednesday Thursday Friday and Saturday according to the same rhythm—this path is easily followed most of the time. But then one day the 'why' arises and everything begins in that weariness tinged with amazement."

—ALBERT CAMUS, *Myth of Sisyphus*, 1955

Life is absurd, and that's okay. No one has written about this more eloquently than Albert Camus in *The Myth of Sisyphus.*[11] The book, a classic in existential literature, derives its title from the legend of Sisyphus, the ancient Greek character who, having defied the gods, is meted an eternal punishment: he's forever doomed to push a boulder up a mountain only to watch it roll down again, and then push it back up, ad infinitum. Camus considered Sisyphus to be a hero of the absurd, a kind of Phil Connors of Greek mythology. Phil Connors is the fictional television weatherman in Punxsutawney, Pennsylvania, from the film *Groundhog Day,* who tried everything, including suicide, to break up the monotony of his mundane existence. Without fail, however, Connors wakes to the same radio song in the same town, destined to follow the same meaningless trajectory of his life. He says, "I was in the Virgin Islands once. I met a girl. We ate lobster, drank piña coladas. At sunset we made love like sea otters. That was a pretty good day. Why couldn't I get that day over and over again?" Who among us can't empathize with this sentiment? Even on a good day, our lives often feel stuck on an endless loop.

Of course, as the author of your own life, you're highly invested in it. Yet occasionally you might awaken to the possibility that from the point of view of the Universe, your life is tiny, accidental, and contains no particular value. The discrepancy between *feeling* that your life is highly valuable with the *knowledge* that you may not be able to justify that feeling is the notion of the absurd. Philosopher Todd May calls it "the confrontation of our need for meaning with the unwillingness of the Universe to yield it to us."[12] It's the quandary that arises when you're unable to articulate why your actions are worth doing or why your life is worth living. This happens when you lose touch with a framework—personal, familial, societal—that could tell you what is genuinely valuable.

This is what has increasingly happened to Western culture. In his classic analysis of American society, *Habits of the Heart,* sociologist Robert Bellah notes how the moral landscape of modern Americans has flattened into the preferences of self-interested individuals. So much so, in fact, that the ultimate goals of a good life have become a "matter of personal choice."[13] People no longer feel they're guided by a solid cultural framework. Instead of *knowing* how to live you feel obliged to *choose* how to live. As Jean-Paul Sartre put it, "everything is permissible if God does not exist."[14]

Consider this: A Gallup World Poll in 2007 surveyed more than 140,000 people from 132 different countries. Among its many questions was: "Do you feel your life has an important purpose or meaning?" When happiness or life satisfaction is examined on a large international scale, researchers typically find the same results again and again: richer nations—as measured by gross domestic product per capita—tend to have happier citizens than poorer nations.[15] The opposite was true, however, when researchers compared answers to this Gallup Poll question. While 91 percent of the people across the world found meaning in their lives, people from wealthier nations like the United Kingdom, Denmark, France, and Japan were most prone to report that their life *lacks* a purpose or meaning, while in poorer nations like Laos, Senegal, and Sierra Leone virtually every-one saw that their life *contains* meaning.[16] The wealthier countries where lack of meaning was more common were also countries with higher suicide rates.

For most of us, existential discomfort is a wave that washes over us quickly yet distinctly, leaving behind the impression, the sense, that perhaps life isn't everything it's cracked up to be, and then the morning alarm clock sounds. It's another day, and you're off to the races yet again. There's a boulder, after all, that needs a good push. But there's another way. It is possible to construct a worldview that can withstand the challenge of the absurd, one that's not only compatible with what modern science tells us about the Universe and humanity's place in it but also retains a sense of justified value, meaningfulness, and sustainable happiness. But first, let's look at the notion of the absurd head-on to better understand how it destroys the illusion of a grand, cosmic sense of meaningfulness. Only then can you begin to take real steps toward personal liberation.

You're
Cosmically
Insignificant,
Impermanent,
and Arbitrary—
and That's
OK

"Our century's revelations of unthinkable largeness and unimaginable smallness, of abysmal stretches of geological time when we were nothing, of supernumerary galaxies and indeterminate subatomic behavior, of a kind of mad mathematical violence at the heart of matter have scorched us deeper than we know."

—JOHN UPDIKE, *Critical Essay on Evolution*, 1985

T he absurd, as noted, refers to how the Universe doesn't yield the kind of meaningfulness you seek from it. A pattern of thought starts innocently enough then digs too deep, rips open the curtains of existence, and you're standing suddenly face-to-face with the absurdity of life. You typically brush up against the absurd via three potential avenues: you come to the realization that life seems (1) insignificant, (2) impermanent, or that (3) all values and goals within it feel arbitrary.[17] Let's look at these three riders of absurdity more closely because a healthy stare at the abyss is necessary to identify the path to the other side of it.

ON INSIGNIFICANCE

If the age of the Universe—some fourteen billion years—was counted in twenty-four hours, it would be fifteen seconds before midnight by the time our species started its evolutionary crawl. Your own life would be over in a fraction of a second. The question of what's significant from the point of view of the Universe can be downright bracing let alone exis-

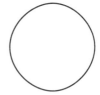

tentially confusing: How are you supposed to figure out what value the cosmos—the planets and the galaxies, filled with their countless twinkling stars and impressive solar systems—assigns anything, never mind your life. As American astrophysicist Neil deGrasse Tyson writes, "The Universe is under no obligation to make sense to you,"[18] a sentiment that would be humorous if it weren't so true.

"Look again at that dot. That's here. That's home. That's us. On it everyone you love, everyone you know, everyone you ever heard of, every human being who ever was, lived out their lives . . . on a mote of dust suspended in a sunbeam."

—CARL SAGAN, 1994, on the last photograph of Earth taken by *Voyager 1* before it left the solar system

It didn't used to be this way: Our ancestors believed Earth was the center of everything. Man was a focal point of God's attention and vice versa. In the creation myths of most cultures, man has had a leading role in the cosmic play of existence. The curse of living in the twenty-first century with an awareness of astrophysics, cosmology, and other sciences is that we know too much. Today, we possess hard, fact-based, scientific knowledge about cosmic proportions and the vast spans of history that predate our own existence that inevitably lead to the conclusion articulated by philosopher Thomas Nagel: "We are tiny specks in the infinite vastness of the Universe."[19]

ON IMPERMANENCE

> "Anyone who has lost something
> they thought was theirs forever
> finally comes to realize that nothing
> really belongs to them."
>
> —PAULO COELHO, *Eleven Minutes,* 2005

As a temporal being, you inhabit a body that ages, gets sick, and will eventually die and disintegrate. Death isn't the sole purview of impermanence, however. The very nature of life itself—our physical, emotional, and intellectual well-being—is temporal. Everything changes and shifts from one moment to the next. Buddhists are especially attuned to the idea of impermanence, *anicca,* which they see as one of the three basic characteristics of existence whereby it's acknowledged that all life is evanescent, in a constant state of flux, and eventually dissolves. You don't have to be Buddhist, however, to grapple with thoughts of impermanence. They are the wormhole of the absurd; it's tempting to conclude nothing's worth the ride if the ride itself is sure to vanish.

ON ARBITRARINESS

> "Be just and if you can't be just,
> be arbitrary."
>
> —WILLIAM S. BURROUGHS, *Naked Lunch,* 1959

The arbitrariness of life revolves around the idea that our aims, goals, and values lack any final justification.[20] We take some life principles and values very seriously—to the degree of letting them guide our choices and actions in life. But are these grand values ultimately justified or mere preferences we've arbitrarily come to endorse? Although we'd much prefer for our values to be somehow grounded in the Universe, we've become increasingly aware that the Universe as such contains no values and has no opinion on ours. Einstein's theory of relativity has nothing to say about why something should have meaning or value. The physical Universe is indifferent.

Life, as a peculiar assemblage of matter capable of self-replicating its form, arbitrarily emerged on the cosmic stage at one point in the history of the Universe. But it generated no objective values. Values are an inherently human invention, and, in fact, the only thing separating human values from animal preferences is that the former are more reflective and can be expressed in language. When you look at ink on paper, you automatically see letters and words. But ink is just ink. The letters exist only in your mind, through your interpretation. It's the same with values. There's nothing behind your values per se. They exist because you and the people around you have endorsed them as such.

More and more people see their life goals and values as something that everyone is free to choose for oneself. But this is worrisome because if all your goals and values are up to you individually, then nothing seems to be ultimately more worth doing than anything else. If a permanent and final justification is needed for your actions to matter, we, as a society, seem to have lost contact with it.

THE POINT OF NO RETURN

Facing the possibility that you are an insignificant, impermanent, and arbitrary being afloat on a pale blue dot in the Universe might sound like a gloomy vision of existence. While you probably don't walk around contemplating the absurd on a daily basis, it's safe to say that thinking about it leaves a lasting impression. As Leo Tolstoy wrote in *Confession,* "We cannot cease to know what we know."[21] Once you've awakened yourself to the possibility that there's no inherent, cosmic value in human life, you can never totally forget it. As there's no going back, the only way is forward. Luckily, there is a way to strive, create, and live joyfully in spite of this knowledge. You have tools at your disposal to create a more meaningful life right here, right now, as the rest of this book will show.

Rather than confront the situation head-on, however, most people choose a work-around and indulge in highly sophisticated methods of distraction. Whole industries have sprung from this well: if you don't want to deal with the absurd, there are a million different ways to entertain and indulge your denial—from selfies and Facebook likes to instantly downloadable entertainment and retail therapy. Accordingly, living with the nagging awareness that life could be cosmically meaningless doesn't often directly translate into an explicit denial of meaningfulness, but more commonly leads to vague feelings of discomfort, defensiveness, and insecurity surrounding your life and your personal goals and values. As long as things are going well in life you might be able to suppress your existential doubts. But when things fall apart—relationships, personal health, or your career—and you would most benefit from having a stable, supportive framework that gives meaning to your suffering, you might become acutely aware of the instability and dimness of your values. This is why distraction is not a good long-term strategy to existential questions.

Of the many distractions our culture has generated to fill the void, perhaps the most prevalent ideology is that you need to be happy. But pursuing happiness contains a paradox, as we'll discover in the next chapter.

HAPPIN

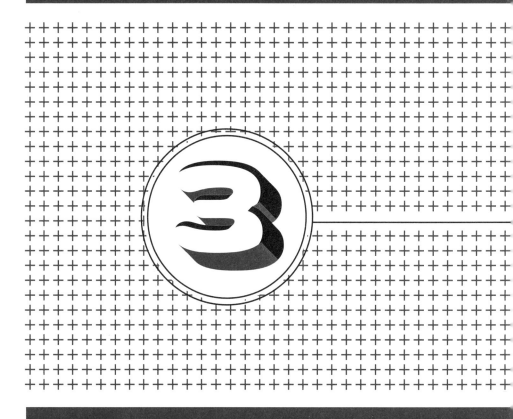

POOR LI

ESS IS A

"Those only are happy (I thought) who have their minds fixed on some object other than their own happiness; on the happiness of others, on the improvement of mankind, even on some art or pursuit, followed not as a means, but as itself an ideal end. Aiming thus at something else, they find happiness by the way."

—JOHN STUART MILL, *Autobiography*, 1873

FE GOAL

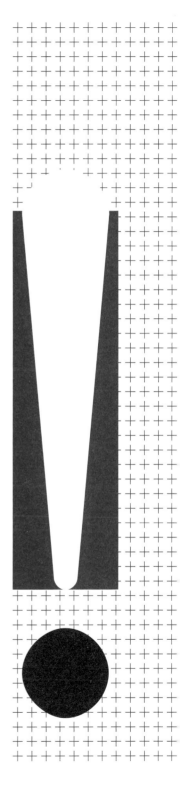

Having lost touch with the kind of grand story that once quenched our ancestors' thirst for meaning, we've psychologized and reduced human existence to a simple model of avoiding pain and seeking pleasure. Happiness has filled the space previously occupied by transcendental values as the one purpose of life worth striving for. Accordingly, happiness has become one of the most celebrated life goals in modern Western culture. It's also big business: while just fifty books were published on the topic in 2000, eight years later that number grew to nearly four thousand.[22] Today, fancy corporations hire chief happiness officers to help ensure employee well-being, and products from soft drinks to perfumes are marketed by a promise of bottled happiness.

Even governments are increasingly paying attention. The *World Happiness Report,* which ranks some 156 countries by how happy their citizens perceive themselves to be, was first released in 2012 and has become a highly anticipated annual event ever since. Since the 1970s the tiny Himalayan kingdom of Bhutan has maintained that the goal of their government is to advance Gross National Happiness not the Gross Domestic Product. Everywhere you look, magazine articles, books, songs, advertising and marketing campaigns, and academic research are dedicated to the subject. It's safe to say that today's egalitarian feel-good-now conception of happiness has become an obsession, the pursuit of which is touted as not only an individual right, but as an individual responsibility.

Derived from the medieval English term *hap*, meaning "luck" or "chance," happiness was originally more about good fortune and things turning out well than an inner state of well-being.[23] From Italian to Swedish, the vast majority of the European words for "happy" originally meant "lucky," including in Finnish, where the word for happiness, *onnellisuus,* stems from the same word as *onnekkuus,* which means "being lucky." The Germans gifted *Glück* to the world, which to this day means both "happiness" and "chance." In the context of these original definitions, happiness was understood as something more like happenstance in that you couldn't control it; it rested in the hands of the gods or with fate or, as the monk in Chaucer's *Canterbury Tales* says, with Fortune: "And thus does Fortune's wheel turn treacherously / And out of happiness bring men to sorrow."[24] Fortune, with a capital *F*, was the work of God's hand, an inexplicable force wholly independent of man's actions and emotional state. This focus on external circumstances reflected a culture where people were much less interested in their inner feelings than ours is today.[25]

During the seventeenth and eighteenth centuries the notion of happiness started to slowly evolve from being about external prosperity to an internal feeling or state of being.[26] When Thomas Jefferson drafted the famous passage in the Declaration of Independence about "life, liberty, and the pursuit of happiness," what he meant by happiness likely still had echoes of prosperity. Since then happiness has more or less come to refer to a positive inner feeling or a tendency to experience one's life in positive terms.

Accompanying this new definition was another key break-through: we came up with the idea that people *ought* to be happy, that happiness was something worth pursuing in life.[27] At first, happiness was considered the goal of society, as codified for example by the United States Declaration of Independence. But especially since the 1960s, Western societies have increasingly perceived happiness as being an individual goal and responsibility. Accordingly, being happy has become a cultural norm and a self-evident aim of life.[28] We want to be happy because our culture tells us that we should be happy. We've acquired a morality where a person's goodness is measured by how good that person feels. Happiness has become the holy cow of our age, an ideal we all ought to strive for.

But here's the thing: happiness is just a feeling.[29] It's an abundance of positive emotions or a general sense of satisfaction with one's life conditions and experiences. And while it's nice to have more pleasant life experiences than unpleasant ones, happiness on its own won't provide a lasting sense of meaning nor is it a way to avoid existential malaise.

In many parts of the world, happiness isn't put on a pedestal. I once had a long discussion about the topic with a Chinese professor of psychology who explained to me that for his parents' generation, being personally happy wasn't a thing. Quite the opposite: being personally unhappy was seen as a badge of honor. It showed the sacrifices one had made for one's family or for the nation. And these sacrifices were perceived as carrying much more value than the fleeting feeling of happiness. Echoing this, a research study from 2004 asked both US and Chinese undergraduate students to write short essays in response to the prompt, "What is happiness?"[30] Many American students emphasized the importance of happiness as a supreme goal in life while, conversely, such strong statements of the value of happiness and its pursuit were totally absent from the Chinese students' writings. So the first thing to note about happiness is that it's not a self-evident goal, and its importance varies from culture to culture.

Second, having happiness as a life goal can easily be counterproductive and may diminish the happiness that you already have. For his book *The Geography of Bliss,* Eric Weiner interviewed a woman named Cynthia who, wanting to settle down, took out a map and decided to calculate where she would be happiest.[31] She wanted to live somewhere with a rich cultural scene, decent food options, and in proximity to nature, preferably mountains. She ended up choosing Asheville, North Carolina, a small but cultural city surrounded by mountains and nature. But when Weiner asked Cynthia if she considered Asheville her home, she hesitated. Asheville was close to meeting all her various criteria but still not optimal. She was still searching. Though she had lived in Asheville for three years, she thought of it as "home for now." Weiner observes how that's "the problem with hedonic floaters like Cynthia and with many of us Americans and our perpetual pursuit of happiness. We may be fairly happy now, but there's always tomorrow and the prospect of a happier place, a happier life. So all options are left on the table. We never fully commit." He continues, writing, "That is, I think, a dangerous thing. We can't love a place, or a person, if we always have one foot out the door."[32]

In their eagerness to derive the maximum happiness out of every life circumstance, the people Weiner interviewed were unable to commit to anything, having lost their ability to enjoy life as it is. And this is by no means the only example of how the pursuit of happiness can be counterproductive. Not only does psychological research demonstrate that people who are most committed to maximizing their own happiness are the ones who are least able to enjoy life,[33] but also that an exclusive focus on one's personal happiness can also damage one's social relationships, which often are the true source of happiness.[34] Finally, the predominant cultural norm that everyone must be happy in fact only makes it harder for us to tolerate life's inevitable, unhappy moments.[35] Feeling unhappy thus becomes a double burden: not only do you feel unhappy, but you also feel guilty for having failed to live up to the cultural norm according to which you ought to be happy all the time.

What if happiness isn't the kind of central life goal we think it is? We value many things—love, friendship, accomplishments, and the ability to express ourselves—not because they bring positive feelings, but because they enrich our lives as such, and we find them worthy by their own accord.[36] The goodness of friendship, for example, can't be reduced to the amount of positive emotions extracted from that friendship. The value of true friendship is especially visible during life's difficult moments when your friend, for example, is seriously ill or going through a crisis and needs support. We value our friends in the bad times, too, knowing that our support is mutually life-enriching even when it isn't all fun and games. Humans are complex; we care about so much more in life than the mere presence or absence of positive feelings.[37] Happiness is a nice experience to have, but making it into a sole life goal is an insult to the richness of what humans actually value in life.

That said, it's difficult to let go of happiness as a goal because our culture is full of messages that remind us that we ought to be happy. Turn on the television—especially during commercials—and there's an entire industry of smiling, healthy, beautiful people selling happiness as a packaged good. Don't be fooled by these false prophets. Don't sacrifice the good things in life in the vain hope of becoming happier. Happiness is just a feeling. That's it. A side product of attaining something valuable rather than the true value itself. Accordingly, the pursuit of personal happiness is a poor answer to the question of what could make our lives truly valuable and meaningful.

Happiness and Heavy Metal: It's Complicated

"I do want to point out that Finland has perhaps the most heavy metal bands in the world per capita, and also ranks high on good governance. I don't know if there's any correlation there."

—PRESIDENT BARACK OBAMA, 2016 Nordic Summit

In both 2018 and 2019, the *World Happiness Report* ranked Finland as the world's happiest country.[38] Whenever overall life satisfaction is measured across the world, Finland, and the other Nordic countries of Sweden, Norway, Denmark, and Iceland, all rank in the top ten, boasting also blue ribbon marks for their relative stability, safety, and freedom. With temperatures that regularly dip below freezing and some towns cloaked in a perpetual state of darkness throughout the long winter, what do the Finns have to be so happy about? Turns out: a lot of heavy metal music.

Heavy metal gets a bad rap but not in Finland. If pop music denotes summers of love, heavy metal is its dark cousin. A country known for its dark and cold winter, Finland has more heavy metal bands per capita than anywhere else on Earth—about sixty-three per one hundred thousand people.[39] In Finland, heavy metal is king, dominating mainstream radio stations and local karaoke bars alike. As one of Finland's bestselling artists of all time, the band Children of Bodom is the king of kings, with sold-out shows everywhere from Helsinki to Rio de Janeiro. Interestingly, being home to a massive number of happy metalheads presents its own set of contradictions that directly relate to data on Finnish happiness and depression.

Based on the Gallup World Poll surveys, the 2019 *World Happiness Report* asked people in 156 countries to "value their lives today on a 0 to 10 scale, with the worst possible life as a 0 and the best possible life as a 10."[40] This is the question to which Finns provide on average the highest scores in the world. Finland's position is actually no surprise at all, because compared to other countries, Finland excels in advancing and promoting the kind of societal factors that we know from research are important for people's sense of life satisfaction: freedom from daily struggle to get bread on the table; extensive social services; freedom from oppression; and trust in government.[41]

However, there's more to happiness than life satisfaction. Some see that it is more about positive emotions. But when you examine how much positive emotion people experience, the table flips, and suddenly countries like Paraguay, Guatemala, and Costa Rica are the happiest places on Earth.[42] Finland falls far from the top, which isn't surprising given the Finns' famous reputation as modest, humble people who don't easily display their emotions. There's an old joke that a Finnish introvert looks at his shoes when he talks to you and a Finnish extrovert looks at your shoes.

Things get even more complicated when we look at the prevalence of depression in different countries. In some comparisons of the per capita prevalence of unipolar depressive disorders, countries like the US and Finland are found close to the top.[43] Although there are significant shortcomings in international comparisons of depression and other research shows that Finland's depression rates are closer to the European average,[44] what is clear is that Finland is far from the top of the world in preventing depression. Paradoxically then, the same country can be ranked high on both life satisfaction and depression.

What this all boils down to is that there is no one thing called happiness. People's emotional lives are complex. Life satisfaction is different from positive emotions, which is different from the absence of negative emotions and depression. If happiness is the prevalence of positive emotions (let alone the displaying of them), Finland is not the happiest country. If happiness is the absence of depression, Finland is not the happiest country. But if happiness is about a general satisfaction with one's life conditions, then Finland, along with other Nordic countries, might very well be the happiest place on Earth.

Furthermore, it would be careless to dismiss the importance of heavy metal music on the Finns' state of well-being. For a country that prides itself on the humble nature of its inhabitants, heavy metal music flies in the face of this characteristic reserve and offers cathartic release. It also offers a channel to express negative feelings, to scream them out rather than trying to suppress them. Actually, that might be more important than we often realize. For one's emotional well-being, it is good to be able to experience a variety of different emotions. Suppressing so-called negative emotions—like the cathartic anger present in many heavy metal songs—is rarely a good idea and often paradoxically leads to lower well-being.[45] A culture repressing and not tolerating expressions of negative emotions is unhealthy and can have a detrimental effect on people's well-being. Accordingly, having a way of expressing the full range of one's emotions is important. And heavy metal might be a great way to scream them out. The question is: If a metalhead screams in a snow-covered forest, will anyone hear him? Whether heard or not, he's probably in better touch with himself and his emotions than his uptight cousin constantly sporting an enforced smile.

IT'S NOT ABOUT THE MONEY

People often make the mistake of equating happiness with financial success. This kind of thinking only services the advertising agencies and corporations that sell the very products you're convinced are the key to personal happiness. Research shows that it's only at the lower end of the income scale that money makes a significant impact in feelings of happiness. People who can't pay their rent or groceries or care for their most basic needs report significantly less well-being than those who can. In these cases, extra income can make a big difference. Once your basic needs are attended to, however, wealth has an increasingly small direct effect on happiness.[46] Several studies have shown that after a certain point on the income ladder, it helps only marginally or not at all, with one recent study even finding that after a certain point, in fact, people's positive emotions and life satisfaction start to decrease.[47]

In North America, this occurs at ninety-five thousand dollars for life satisfaction and at sixty thousand dollars for positive emotions. In Western Europe, the turning point is at one hundred thousand dollars/fifty thousand dollars while in Eastern Europe, it's as low as forty-five thousand dollars for life satisfaction and thirty-five thousand dollars for positive emotions. Furthermore, although many industrialized nations have made significant financial gains, this hasn't translated to any more happiness, with American social psychologist Jonathan Haidt summarizing the results, writing, "As the level of wealth has doubled or tripled in the last fifty years in many industrialized nations, the levels of happiness and satisfaction with life that people report have not changed, and depression has actually become more common."[48] Once people adapt to a new baseline of wealth, their initial happiness dissipates; the new creature comforts become the norm and, over time, are taken for granted at least until the next latest and greatest tech device or luxury item is released on the market, and everyone swivels their attention toward its pursuit.

On a surface level, most of us can probably reject consumerism and materialism as a life goal. When asked, we usually report something grander as a motivating source. But beneath the surface, it's often another story. Although we might be reluctant to admit it, many of us are addicted to the hedonic treadmill where the promise of happiness is always a bit out of reach. As Chuck Palahniuk writes in *Fight Club,* "You have a class of young strong men and women, and they want to give their lives to something. Advertising has these people chasing cars and clothes they don't need. Generations have been working in jobs they hate, just so they can buy what they don't really need."[49]

The advertising industry is a two billion-dollar[50] propaganda machine that has but one goal: to make your current life feel inadequate. To make you feel that what you have now isn't enough. Consumerism stops the moment you become satisfied with your life, when you're able to say, "I don't need anything. I already have everything that I want." This is the kind of state that many religious doctrines from Christianity to Buddhism try to guide us toward. But in this secular age, the belief is that it is better to spend billions of dollars on messages that prevent anyone from reaching that state.

Today more than ever before, we're presented with so many different choices of competing products it's tempting to get caught up in the happiness trap. We prize the ideas of freedom and choice even when we know that too much of a good thing can lead to harm or to addictive behavior. It's an irony of modern life that the more choices there are, the less likely we feel confident in our choice. If you can avoid making a choice, you are more likely to do so. Psychologist Barry Schwartz refers to this idiosyncrasy as the "paradox of choice": we value choice and crave it despite the fact that too many options and choices can undermine our sense of happiness.[51] Our ancestors didn't contend with this dilemma to the extent that we do now. Starving was much more common than having too many enticing food options to choose from. The best way to manage the bombardment of daily choices is to become what Schwartz, following Nobel Prize–winning economist Herbert Simon, calls a satisfier, that is, evaluate your choices, pick one that's satisfactory or "good enough," and move on with your life.[52] Stop trying to maximize every detail of every purchase or decision—that just leads to more stress, regret, and dissatisfaction.[53] You've got better things to do with your time, energy, and resources. But to fight the powerful influence that the constant bombardment of advertising has on your life ideals, you need an even more powerful inner compass. You need to have some self-chosen values and life goals that are so strong and salient that you can retain your integrity even in an advertisement-filled society. For that, having a good grasp of what makes your life meaningful can be of significant help; behind all the surface glitter, and the presence or absence of the latest expensive gimmicks, your life probably already contains much of what can make it meaningful.

4

"Be not afraid of life.
Believe that life *is*
worth living, and
your belief will help
create the fact."

—WILLIAM JAMES,
Is Life Worth Living, 1897

· YOUR LIFE
IS ALREADY
NGFUL

Most people in most circumstances experience their lives as being pretty meaningful no matter the lingering existential doubts. When I told professor Laura King, one of the leading experts of psychological research on life's meaningfulness, that I was in the process of writing a book on the subject, she gently pulled me aside in the corridor of Portland's Convention Center and issued a sage warning: don't tell people that their lives are meaningless—that would be irresponsible because decades' worth of research demonstrates the opposite. In an influential 2014 article published in *American Psychologist,* King and Rutgers's professor Samantha Heintzelman reviewed various nationally representative surveys and other evidence to see how much meaningfulness people experience on average.[54] Turns out it's a lot. When a broad survey of Americans over the age of fifty were asked whether they felt their life had meaning, 95 percent said yes.[55] Another survey asked a large national sample of people to rate their agreement with specific statements like "My life has a clear sense of purpose" on a scale from 1 (not true at all) to 5 (completely true). The average was quite high, 3.8.[56] And this trend extends outside the US, too. As noted before, of the 140,000 people from 132 nations surveyed by the Gallup World Poll about whether their "life has an important purpose or meaning," 91 percent answered yes, with the percentage even higher in some of the world's poorest countries.[57] Other studies show that people facing various health problems, like fighting cancer, typically still find their life meaningful.[58] King and Heintzelman write, "Evidence from large representative samples and the body of research using an older and a newer measure of meaning in life strongly point to the same conclusion: Life is pretty meaningful."[59]

Despite the absurdity of existence and the fact that life is, in general, cosmically insignificant, impermanent, and arbitrary, most people most of the time seem to experience their lives as being meaningful. Should we say that most people most of the time are mistaken? Perhaps it's our duty to reveal the gloomy existential facts to them? This inherent paradox is often the fork in the road for philosophers and psychologists. Certain philosophers want to argue that people are mistaken in holding their lives in high regard and the absurdity of existence should be rubbed in their faces. Psychologists, instead, tend to take people's evaluations at face value: if a person feels that one's life is meaningful, then that life is meaningful indeed. While I side with Professor Laura King and the psychologists, and believe that, for the most part, we should believe people who say their lives are meaningful, it's also important to get to the root of this very human paradox.[60]

The paradox between life being absurd and the fact that people still experience high levels of meaning in their lives is, to a significant degree, the result of us not really understanding the question of meaning correctly. More accurately, we confuse two separate questions. One of them we no longer seem to find an answer to, and the confrontation with this question typically leads to an existential crisis. The other question, however, still has a solid and life-affirming answer to it through which to experience meaning in life. What we need to realize, however, is that the former question—"What is the meaning of life?"—is, in fact, a historical by-product of Western thinking that's only emerged in the past couple of centuries. Times have changed, but we still seek a kind of meaning that only makes sense in the old, abandoned worldview. Our current crisis of meaning is thus an understandable mistake given the intellectual history of Western society. But it's still a mistake in need of correction.

We'll start making that correction in the next chapter. To do so, however, we first need to understand why humans search for meaning in the first place.

REFLECTIVITY CURSED HUMANS WITH THE NEED FOR MEANING

"Humans may resemble many other creatures in their striving for happiness, but the quest for meaning is a key part of what makes us human, and uniquely so."

—ROY BAUMEISTER,
"Some Key Differences Between a Happy Life
and a Meaningful Life," 2013

Among the key physical peculiarities that make humans stand out even among our closest primate cousins is the size of our brain compared to our body. Our ancestors some two million years ago had brains of about 24 to 37 cubic inches. Modern humans typically have brain sizes of approximately 73 to 79 cubic inches.[61] This dramatic growth spurt, called the cognitive revolution by scientists or the "Tree of Knowledge mutation" by author Yuval Noah Harari,[62] ultimately separated the upright walking naked ape from his animal cousins. There are many theories about what led to this rapid arms race in brain size in the human species and what unique abilities this new processing power enabled—language, cooperation, culture, religion, and so forth—but let's concentrate on this key feature: humans became reflective.

Reflection is the ability to take a third-person's look at one's own life. Instead of immediately responding or reacting to whatever happens in the present moment, we're able to step outside of the situation to stop and think. We can contemplate our past actions and make predictions about the future while simultaneously consolidating both sets of information to make conscious decisions about how to behave in the present.

Our capacity for reflection is a uniquely human tool and sets us apart from other animals trapped in the present moment, unable to plan for the next month let alone the next decade of their lives.[63] Humans have basic animal instincts, but we also have the brain power to override such impulses to focus instead, for example, on long-term goals, the rewards of which will only be reaped days, months, or years from now. Reflection allows for planning, collective action, and long-term goal setting, which in turn has enabled us to create artwork, architecture, and tools, to name a few things, unimaginable for any other animal. It took approximately two hundred years to build Notre-Dame in Paris. Such a monument is testimony to the creative potential totally unseen on this planet before human existence. Reflection isn't just about planning for the future, and carrying out epic-sized projects, however. Reflection also lets us connect with the past, which can further enhance the meaningfulness of our lives. Philosopher Antti Kauppinen has argued that "building on the past gives life a kind of progressive narrative shape,"[64] which sounds more meaningful than a life composed of isolated episodes. Neuropsychology agrees.

A neuroimaging study carried out at Northwestern University with eighty-four participants demonstrated that increased connectivity in the medial temporal lobe network, which is known to be implicated in mental time travel to experience the past or the future, is correlated with people's reports about how much meaningfulness they experience in life.[65] A trip down memory lane can thus provide a nostalgic sense of meaningfulness. As humans, we're also peculiarly calibrated to be able to find meaning in the future. Reflection makes hope possible: we can envision a better world and make plans to actualize it. Valuable goals in the future are often what make present-day efforts, and even struggling, meaningful. We can endure present-day suffering and pain much better if we're able to retain the hope and the belief that something valuable awaits us in the future.

But there's a price to pay for all this capability for reflection. Because of it, we don't settle for the instinctual goals that drive most other animals. It's both a blessing and a curse that we're inescapably trapped in a world that extends both backward to the past and forward to the future. We plan and worry about things that might happen or might not happen in the distant future. We mull things over from our past, rehashing old wounds or cherishing memories. Our problem—as compared to most animals—is that we can stop in the middle of an activity and, with a kind of reflexive self-reflection, ask ourselves: What's the point? Why am I doing this?

Reflection, thus, creates the need for justification.[66] When the question of why arises, we need a satisfactory answer. We need to be able to endorse our actions even after we've reflected on them. Herein lies the origin for our need for meaning. As reflective animals, we need to feel that our activities have a point, a reason or a purpose behind them, and that they—somehow—matter, and contribute to something worth contributing to.[67] In order to answer ourselves, we need a framework, some kind of worldview, that tells us specifically which activities and goals are worth doing and which ones are pointless; in short, when we come to that inevitable fork in the road, we want a meaningful worldview to guide us toward the path worth taking. Lacking such a framework of meaning can have grave consequences.

Already during the Second World War, Erich Fromm, the social psychologist and psychoanalyst, observed that modern man is "freed from the bonds of pre-individualistic society" that had previously limited him.[68] While many celebrated this liberation as the final step toward man's evolution as a self-standing individual capable of self-actualization, things didn't go as planned. Despite being restricting at times, the traditional cultural framework gave humans a sense of security, comprehensibility, direction, and significance. In other words, it had provided humans with a robust sense of meaning in life. Absent that framework, however, people still needed to know what to do with their lives and how to make them worth living. Unfortunately, the newly liberated culture didn't have suitable or comforting answers, which unsettled many people, making them feel isolated, anxious, and at a loss of direction. The liberation that was supposed to happen was instead subverted to an escape from freedom, with people submitting to whatever authority was willing to give them firm answers to life's big questions and thus the stability they so desperately needed.

According to Fromm's analysis, modern man is "anxious and tempted to surrender his freedom to dictators of all kinds, or to lose it by transforming himself into a small cog in the machine, well fed, well clothed, yet not a free man but an automaton."[69] For Fromm, this was one of the root causes of the rise of fascism in Europe in the 1930s, the atrocities of which are all too well known. There are alarming similarities across the political landscape of the Western world today, which makes it imperative that we create a reflective framework and value system that not only feels universally justified but is also able to withstand the cynicism and divisiveness of our modern times. The other option is uncertainty and regressive, authoritarian value frameworks, which undermine the principles of care, equality, and freedom at the heart of the lifestyle we have built for ourselves since the dawn of the first true democracies in the eighteenth century.

If we can't do that, we've wasted a perfectly good cognitive revolution.

The Art of Simultaneously Searching for and Having Meaning in Life

Why search for meaning if our lives are already meaningful? This simple question reveals the Western bias in our thinking about meaningfulness. Professor Michael Steger, another key expert on the psychology of the meaning in life, examined people's *search* for meaning in life and the *presence* of meaning in life and found that the results were opposed in the US: the more there was presence of meaning in a person's life, the less prone that person was to pursue more.[70] We operate from a deficit mode as regards meaning in life: we get interested in the topic mainly when it's lacking. However, when Steger investigated the same matter in Japan, he discovered that there the relation between the two was not contradictory but harmonious: a person already living a meaningful life was more prone to reflect on how to live even more meaningfully. It might be that this openness to reflect on the meaning of one's life is what led the individuals to make life choices that increased their sense of meaningfulness in the first place.

Eastern cultures might be wiser than the West in this regard. Although you may already have some presence of meaning in your life, it's possible and often life-enhancing to search for even better sources of meaningfulness. This is not a desperate attempt to fill a void, but an invitation to reflect on your life to find more ways of living that work in harmony with the activities, choices, and relationships that could make your days more meaningful. We are works in progress and the joy of being human is that we know this—we fundamentally know there's room, in each of us, for more understanding, improvement, and personal fulfillment. Rather than bemoaning what you may lack, take stock of what you already have and find ways to build on it. Free yourself to understand reflectively what you already may grasp intuitively, that your life is already meaningful.

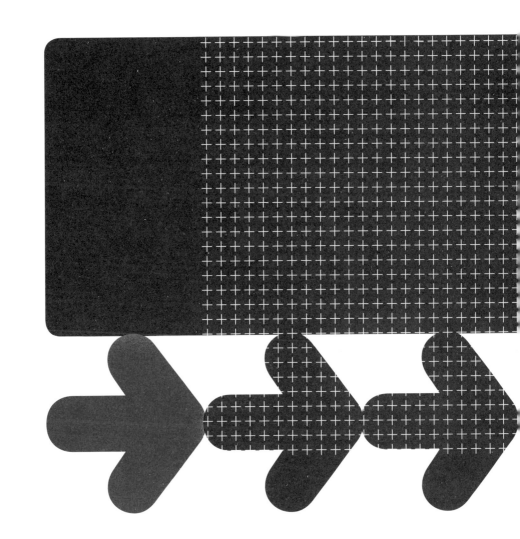

PART TWO: The Question

of Meaning: A New Perspective

Your Existential Crisis Makes You MOD

"What we understand as 'atheism' would have been unintelligible to the classical mind. Certainly, there were

ERN

disagreements on the nature of the gods or their activities, and sometimes even the denial of the existence of certain gods. But the notion, intrinsic to the modern understanding of atheism, of immanence—of the world existing quite free of any sort of transcendent realm—would have been almost unintelligible to them."

—GAVIN HYMAN, *A Short History of Atheism*, 2010

Imagine this: You're charging your iPhone at the airport, and a man approaches you and asks, "Do you believe in electricity?" You and I, and I'd wager every other person in the airport, know that the whole of today's modern lifestyle revolves around electricity and, as such, there's no reason to debate belief. The question itself is nonsensical. The man leans in for a follow-up, asking, "Do you believe in God?" This question carries a bit more weight and, no matter your answer, it's likely something you've previously contemplated or even debated before in your life. Unlike the nonsensical question about electricity, you understand the question of your belief in God as being a genuine query. Your understanding of this religious questioning makes you modern.

To Europeans living five hundred years ago, the question about God would have been just as peculiar as the question about electricity is to us today. God's presence—not electricity—was everywhere. Theirs was a world dominated by the supernatural—spirits, demons, and magic. It was commonly believed that someone of ill health was possessed by demons who could drive that person to commit malevolent acts.[71] Relics of saints held healing power. Storms, droughts, plagues, and periods of fertility were seen as acts of God. People engaged in regular collective rituals, like the reading of the Gospels in cornfields to ward off wicked spirits who could harm the harvest.[72] In the words of German sociologist Max Weber, the world of the premodern people was *enchanted*.[73] The existence of God and spirits wasn't a question of belief, but an immediate certainty. The whole cosmos was a meaningful entity in which all parts hung together in a purposeful plan. This was true not only of medieval Europe but also globally. Of course, the names and functions of the various spirits varied from culture to culture. Some cultures believed in an omnipotent creator God, others in a myriad of local spirits.[74] No matter the particularities of one's beliefs, however, the enchanted world was one in which various spirits, demons, gods, and cosmic forces constantly influenced everyday occurrences big and small.

In this world of enchantment, there was no clear distinction between natural and supernatural explanations; no scientific worldview had yet been discovered or developed to ground the former. In theory, one could deny or debate the existence of individual spiritual beings—whether a certain spirit existed or God's exact nature and powers—but if one wanted to stop believing in the whole enchanted worldview, there simply wasn't a Plan B. No alternative worldview existed. Disenchantment—as a possible worldview—had not yet been invented. The conceptual tools and ideas upon which to base not believing simply didn't exist. Instead, enchantment formed the totality of one's worldview and, as such, it was impossible to abandon the collective rituals built into one's day-to-day life that supported and strengthened this ideology.

Because of our fundamental difference in worldviews, the manner in which modern people talk about the meaning of life wouldn't have made sense to medieval peasants nor for that matter to great ancient thinkers like Aristotle or Epictetus. For most of human history, people didn't question the meaning of life because there was no need to think about it. In an enchanted cosmos, it was obvious that all life existed to fulfill some larger, cosmically driven or divinely given purpose. Prior to modernity, our ancestors' views of the cosmos and man's place in it was positively quaint. Their world, at least cosmically speaking, was orderly in ways ours certainly is not. The ancient Greeks knew nothing about the secret lives of black holes to say nothing about postmodern art or brain scanners. Even the boldest thoughts of Aristotle, the famous Greek philosopher who lived in the fourth century BCE, were colored by the enchanted worldview within which he and his contemporaries lived.

If there was a contest to decide which single individual has most influenced Western thinking, Aristotle would be in the semifinals, facing heavyweights like Jesus and Sir Isaac Newton. In one of the most celebrated and studied books on ethics ever written, *The Nicomachean Ethics,* Aristotle pondered the idea of the highest human good. More specifically, he was looking for "some end of our actions that we wish for on account of itself."[75] He aimed to unravel what makes us special as compared to animals, believing that our human nature itself holds the clues for what determines the highest human good. It's tempting to conflate this idea of the highest human good with meaning. I would argue, however, that Aristotle wasn't discussing the meaning of life when he investigated the concept of the human good. He surely discussed the idea of human purpose, but not in the right way. Or, rather, his investigation was limited by the era in which he lived, and it lacked an awareness of one central element: the absurd.

In much the same way that we take electricity for granted, Aristotle didn't doubt the existence of a cosmic order, so much so, in fact, that it never occurred to him that such an order may not exist. The enchanted world was a meaningful whole and, like every other creature, human beings had some inherent purpose or virtue, the fulfillment of which defined the human good: the virtue of a horse is to run and carry the rider; the virtue of an eye is to give us sight; and thus for Aristotle, there must exist "the virtue of a human being," some form of excellence unique to being human.[76] Observing that our capacity for rational thinking sets us apart from other animals, he concluded that the human good must be about living in accordance with this rational soul, which requires certain virtues. For him it was never a question of whether there *is* or *isn't* a purpose to human existence. In the enchanted cosmos Aristotle lived in, it was self-evident that humans had a purpose since everything had some inherent purpose; it was only a matter of us discovering it.

The grand question about life for Aristotle and for the next few millennia of Western thinking was about the end of man. This question, called *telos* by the ancient Greeks and *summum bonum* by the medieval Christian thinkers, was the focus of Western thinkers until modernity, and sought to address the ultimate end of humankind, the essential *why* of our existence. It was a question about *what human beings are for*, in the same sense we might ask what bicycles or knives are for; riding and cutting, respectively. What united both Greek and Christian thinkers from Aristotle to Thomas Aquinas was the fact that they never questioned the possibility that humanity had an end. Theirs was a worldview where the cosmos was intelligible and human beings were created for a purpose. So the task of the thinker was only to unveil and discover the human good or the human end already present. As Professor Joshua Hochschild argues, the end of man was "the question about human life asked for most of Western history." [77]

Starting somewhere in the seventeenth century, however, a more scientifically driven worldview started to slowly gain prominence in Western societies. This new worldview first carved out a separation between the natural and the supernatural and then started to push the latter into the margins. In the span of a few centuries, science killed the enchantment of the Universe. There were other culprits as well—the rise of humanism and individualism, urbanization, increased mobility, industrialization, democracy, and more bureaucratic governments—but the scientific worldview was most influential in turning the *Enchanted Cosmos* of premodern times into a seemingly disenchanted, meaningless, and *Mechanical Universe*. In the Enchanted Cosmos, the question about the end of man made sense, but it didn't fit into the Mechanical Universe where humanity no longer had a self-evident place in the grand order of things. This led to the need to ask a new kind of grand question about life. In 1834, a man named Thomas Carlyle hinted at the seemingly simple question "What is the meaning of life?" and we, as a society, have been grappling with the existential fallout ever since.

A MEANINGFUL INVENTION: CARLYLE'S BIG QUESTION

"Rightly viewed, no meanest object is insignificant; all objects are as windows, through which the philosophic eye looks into Infinitude itself."

—THOMAS CARLYLE, *Sartor Resartus*, 1834

While it can't be said that Thomas Carlyle, the Victorian-era Scottish essayist, satirist, and historian was the only man to ponder the meaning of life, he was the first to write about it in the English-speaking world. Published between 1833 and 1834, Carlyle's book *Sartor Resartus* was notable for several reasons: Ralph Waldo Emerson wrote its preface, Herman Melville and Walt Whitman cited it as a key influence on *Moby-Dick* and *Song of Myself,* respectively, and today it's often cited as *the* book to mark the transition in English-speaking literature from the Romantic to the Victorian period.[78] It also contains the earliest known writing in the English language of the phrase "meaning of life."[79]

Sartor Resartus was written at a particularly tumultuous time in world history when nearly every aspect of daily life was affected by any one of several revolutions happening worldwide: the French Revolution altered the political world and its aftershocks were still being felt across Europe; the Romantic revolution fostered emotions, self-inspection, and introspection; nearly every aspect of daily life was transformed by the Industrial Revolution; and the scientific revolution threatened the religious worldview. Carlyle's text opens with the following lines: "Considering our present advanced state of culture, and how the Torch of Science has now been brandished and borne about . . . ; how, in these times especially, not only the Torch still burns, and perhaps more fiercely than ever, but innumerable Rush-lights and Sulphur-matches, kindled thereat, are also glancing in every direction, so that not the smallest cranny or doghole in Nature or Art can remain unilluminated."[80]

The "Torch of Science" burning so fiercely that no "cranny or doghole" can remain unilluminated—one could not better describe the sheer intellectual power with which scientific thinking forced itself into people's lives and reshaped their dearly held truths and worldview. What was previously taken as self-evident and beyond question—that the world is a meaningful whole and humanity has a special part to play in the unfolding of that world—suddenly lost its foundation. Perhaps, then, it wasn't shocking that readers should find the book's protagonist, middle-aged Professor Teufelsdröckh of Weissnichtwo, overwhelmed by the meaninglessness of his trivial life. His is a depression that often accompanies a major life transition but one that's made especially acute by the uprootedness many people felt in this era of the nonstop march of industrialism and other transformations. Unlike in previous times, religion and tradition no longer seemed to hold all the answers.

In the novel, which also serves as an allegory of Carlyle's own search for meaning, we're shown how "Rational University," that is, an increasingly secular world that is "in the highest degree hostile to Mysticism," infected Teufelsdröckh with religious doubt, causing him to question his faith and the very existence of God.[81] Doubt darkens into what he calls "the nightmare, Unbelief," and the professor soon finds himself seemingly alone in a cold and silent world, writing: "To me, the Universe was all void of Life, of Purpose, of Violation, even of Hostility: it was one huge, dead, immeasurable Steam-engine rolling on, in its dead indifference, to grind me limb from limb." Bereft of belief and left to his own philosophical devices, he declares—using the phrase the "meaning of life" for the first time—"Our Life is compassed round with Necessity; yet is the meaning of Life itself no other than Freedom, than Voluntary Force: thus have we warfare; in the beginning, especially, a hard-fought battle." The essential human battle for him is thus between Necessity and Freedom: a man either bound by appetites, bodily desires, and other earthly matters or a man dedicated to transcending such things in order to follow a higher moral duty in his work. For Carlyle, this is the meaning of life: by engaging in purposeful work, we can transform our personal ideals into reality and attain a real sense of fulfillment. He writes, "Work while it is called To-day, for the Night cometh wherein no man can work."

It's comforting to know that the meaning of life, far from being a burning question that's teased humans since the dawn of time, was a phrase coined less than two hundred years ago by an author whose semiautobiographical protagonist is himself the author of a book entitled *Clothes: Their Origin and Influence*. Considered a professor of "things in general," Teufelsdröckh expounds on exactly that in his sartorial-inspired tome, including the importance of the "Clothes-wearing man," the Dandy, and the proper German way to wear a collar: low behind and slightly rolled.[82] While *Clothes* is seemingly innocuous, it's also a poioumenon, a type of metafiction that gave Carlyle ample room to voice more weighty, philosophical concerns. Yet, despite doubting the existence of any value in the modern world, Carlyle clearly didn't think it was all gloom and doom, and he imbued the text with words of hope and the conviction that man could, indeed, traverse the existential wilderness and emerge from it triumphant. Still, the whole book was in a sense a symptom of the fact that Carlyle had lost touch with the stern Calvinist religious faith that his parents had enjoyed. *Sartor Resartus* can be read as his struggle to grapple with this loss of faith while living in what he calls "an Atheistic Century"; it's his attempt to come up with a way of understanding life that's compatible with a loss of faith in traditional Christianity.

Often named as one of the most influential public intellectuals of the nineteenth century, Carlyle inspired a host of thinkers; everyone writing or thinking about meaning and existential crisis in the English-speaking world was, in one way or another, reacting to his work. At the same time in continental Europe, philosophers like Søren Kierkegaard and Arthur Schopenhauer picked up where Carlyle dropped the first stitch, with Kierkegaard writing, "What, if anything, is the meaning of this life?" in *Either/Or*, a seminal early work from 1843.[83] An existential fever seemed to take hold in learned circles, sweeping up everyone from philosophers and novelists like Ralph Waldo Emerson, Samuel Beckett, George Eliot, and Leo Tolstoy, composers like Richard Wagner, and biologists like Thomas Huxley (known by his nickname "Darwin's Bulldog"), with Schopenhauer helping to lead the charge. In his essay "Human Nature," he asks point-blank: "What is the meaning of life at all? To what purpose is it played, this farce in which everything that is essential is irrevocably fixed and determined?"[84] Tolstoy's *Anna Karenina*, published in 1878, popularized the idea of existential malaise to the general public—previously it had mainly been the purview of a select circle of intellectuals. Even before the publication of his masterpiece, Tolstoy had grappled with a disenchantment similar to Carlyle: like the Scottish philosopher, Tolstoy struggled to come to terms with the scientific worldview. It seems no accident that a few months before writing in his diary that "life on Earth has nothing to give," he had been reading about physics and pondering the concepts of gravity, heat, and how a "column of air exerts pressures."[85] In understanding more about the mechanistic laws of nature, he lost his faith in the transcendent, writing, "Far from finding what I wanted, I became convinced that all who like myself had sought in knowledge for the meaning of life had found nothing."[86]

Tolstoy, along with Carlyle and Schopenhauer and other contemporaries, was among the first to realize the full implications of the new scientific worldview: it reduces humanity into a biological organism that has no inherent purpose, good, or value. As Tolstoy put it: "You are a temporal, accidental conglomeration of particles. The interrelation, the change of these particles, produces in you that which you call life. This congeries will last for some time; then the interaction of these particles will cease, and that which you call life and all your questions will come to an end. You are an accidentally cohering globule of something. The globule is fermenting."[87]

Of course, knowing something and liking its discovery are two different things. Many didn't like the inconvenient truth that lay at the heart of the scientific worldview, but that didn't prevent the worldview from spreading. By the end of the twentieth century, the general public had withstood generations of existential malaise, so much so, in fact, that the seemingly unanswerable Big Question—What is the meaning of life?—felt less like a man-made invention and more like man's eternal struggle. It undoubtedly made some long for simpler, more enchanted times or, barring that, at least some prudent advice on how best to style one's collar.

HOW SCIENCE GREW OUT OF RELIGION

> "This most beautiful System of the Sun, Planets, and Comets, could only proceed from the counsel and dominion of an intelligent and powerful being."
>
> —SIR ISAAC NEWTON, *General Scholium*, 1713

Although nowadays it's commonplace to pit science and religion against each other, this wasn't originally the case. For several centuries, the scientific revolution progressed within the context of Christianity and a firm belief in God, rather than opposed to it. Indeed, the rational and logical analysis that science relied on was first developed within the context of theological studies as a way to better understand God and the world he created.[88] Initially, scientific investigations were a way of celebrating and getting closer to God. Given a rational cosmos designed by divine will, scientists like Newton were simply deciphering the language of God. The goal was to better understand the intelligent heavenly plan behind the Universe. Johannes Kepler, the seventeenth-century German mathematician and astronomer whose laws of planetary motion were a major force in the scientific revolution, was motivated by a wish to demonstrate that God had created his Universe on the basis of geometry. Kepler transformed from being a theologian to an astronomer when he realized that "God is also glorified in astronomy, through my efforts."[89]

The scientific worldview thus started as a spin-off of the Christian worldview but it soon suffocated its parent. More and more thinkers began to understand how the various elements of this new worldview were increasingly not dependent on God but were instead able to stand on their own. While the word *atheism* first appears in the English language in 1540,[90] it took some time before its definition was narrowed down from denoting general heresy to meaning an outright denial of theism. Like anything that poses an affront to the status quo, the word was first used accusatorily; most people saw atheism as being on par with witchcraft and sorcery. It wasn't until the mid-eighteenth century that Frenchman Denis Diderot became the first self-confessed, public atheist philosopher.[91] As a philosophical trend, atheism moved like wildfire . . . at least in certain circles. By the latter part of the nineteenth century, Tolstoy observed that, among the learned circles of the Russian and European elites, "hardly one in a thousand professed to be a believer."[92] By the late-nineteenth century, many universities had dismissed the previously accepted thinking of religious dogma, relegating religious argumentation to a marginal role. Religious thinkers and believers, of course, didn't disappear, but the practice of belief and faith began to turn inward. Belief was becoming more of a private matter. In the public sphere, certainly in politics and in the workplace, religious believers were expected to participate in rational discussions wherein the supernatural could no longer be enlisted as part of one's argument. When making decisions on when to harvest one's crops or whether to build a dam, one relied on scientifically tested knowledge, not divine revelation, and consulted experts, not spirits.

To be sure, enchantment, religion, and the supernatural still impact many people's lives today—not as the self-evident background tapestry of our worldview, however, but as something that coexists with the scientific worldview in more or less tension. Religious believers today must navigate the space between their private, more enchanted beliefs and the disenchanted, rational, and modern way of seeing the world. And while this disenchanted worldview has led to many gains in technology and how to run a society, it's also revealed a terrible possibility: What if the Universe didn't create human beings to fulfill a grand purpose in life? What if the meaning of life is that there is no meaning?

Thus, the question "What is the meaning of life?" is first and foremost reactionary. It was invented as a consequence of the spreading of the scientific worldview and the consequent disenchantment of the world. The old self-evident purposefulness of the whole cosmos, including human beings within it, was challenged. In this context, it became acutely important to ask for what was lost. And a phrase was invented to describe what we used to have: meaning of life. But let's not lay all the blame on science for our existential crisis. After the scientific worldview entered human consciousness, someone needed to invent the idea that we ought to experience our lives as meaningful. Enter the Romantics.

A ROMANTIC

TION

"Trying to work out the meaning of life can be rather like trying to assemble Ikea furniture when you're convinced that you're missing a piece or haven't been given proper instructions. But the real problem is that you're trying to put together an elaborate Maråker cabinet when you have only got a standard three-shelf Billy bookcase. Something only seems to be missing because you're expecting much more."

—JULIAN BAGGINI,
Revealed: The Meaning of Life, 2004

In addition to playing a major role in launching the modern quest for the meaning of life in the Western world, Thomas Carlyle, Arthur Schopenhauer, and Søren Kierkegaard had something else in common: German Romanticism. Carlyle translated the writings of many German Romantics into English and wrote *The Life of Friedrich Schiller,* a book about one of the leading German poets of the late 1700s. As a German himself, Schopenhauer was familiar with much of the work of his compatriots and, in a sense, his pessimistic philosophy was a reaction to them. He certainly tempted intellectual conclusions the Romantics didn't dare. Kierkegaard moved to Berlin to listen to the lectures of Friedrich Schelling—another German Romantic—and wrote much of his first existential treatise, *Either/Or,* while there. The historical detective work on the shared intellectual background of these three men leads to a question: Was there something in the German idealism that planted the seeds for the existential crisis all three men experienced?

One of the central figures of early German Romanticism was a poet named Georg Philipp

Friedrich von Hardenberg, more commonly known as Novalis, whose one true love in life—Sophie von Kühn—died after their engagement but before they were able to marry. Novalis found himself love- and grief-stricken and full of idealistic visions about love and life, which he translated into poems and other writings before dying of tuberculosis at the age of twenty-eight. As a direct reaction to Europe's increasingly rationalized and disenchanted worldview that was swiftly secularizing the spiritual and the sacred, German Romanticism turned its focus inward: enchantment should come from within. Novalis, along with other like-minded Romantic poets-cum-philosophers, championed human emotions, elevating them to almost sacred status. They worshipped love and emotional authenticity and believed that they should guide our lives.

Today, whenever a friend or a loved one is in the midst of contemplating a big life decision, it's commonplace to say, "Follow your heart." This advice is very much a Romantic invention. Before the Romantics, it was more common to simply soldier on—ignore your heart and fulfill your duty. For the Romantics, however, following one's heart was less a sentiment and more a directive: the Romantic bravely disregarded social norms, paternal expectations, rational advice, and other limitations to fulfill his heart's calling. The ultimate hero was a poet in the throes of unrequited love totally careless about the duties and practicalities of life, preferably dying an untimely death when his fragile body or heart betrayed him. Despite knowing he'll never win the affection of his love interest—or, worse, like Novalis, it may be tragically lost—he remains dedicated to his love and uses poetry to express his feelings.

Romanticism thus initiated an idea that since has been promoted by Hollywood movies and countless pop songs: love should be a head-over-heels experience. True love is waiting for you somewhere out there, and when you meet him or her, it will be love at first sight, an immediate certainty that this person is your heart's chosen one and you'll love each other forever after. Besides Romanticism promoting an unrealistic expectation of love, which "has been a disaster for our relationships," and has had "a devastating impact on the ability of ordinary people to lead successful emotional lives" as philosopher Alain de Botton has argued,[93] it also transformed the ideal into an imperative: you should never settle for less, a statement that sounds good but rarely works in the reality of managing the myriad complex emotional components of a romantic relationship let alone the mundane problems of cohabiting.

This line of thinking carried over into work, too: don't settle for any other job; find your true calling. Sound familiar? How many times have you been told—in advertisements, films, songs, self-help books—that you shouldn't settle in love, work, life, or happiness? Implicit in this directive is the idea that all of us, somewhere, somehow, have an inner calling. All we have to do is find it, and then we'll understand what it is we were put on Earth to do. What the Romantics did was take the Christian idea of a calling—as in being called by God to do a certain kind of work—and replaced "God" with "heart." So, there was still one true mission in life that you were meant to do, but instead of it being given to you by God, it was hidden inside you all along. This idea became a kind of motto for self-help literature—find your true calling—and, sadly, doesn't have much to do with reality.

And here's how it all connects to the meaning of life: in advocating for a heart's calling, the Romantics were, in essence, promoting the idea that we're each promised a meaningful life; we just have to discover it. By following your heart, a mission will emerge wherein the wholeness of your life will be revealed and suddenly fall into place, and your life will be overwhelmed with clarity, certainty, and a sense of meaningfulness.

The Romantics promised us roses, but reality often doesn't smell so sweet. I would argue that the combination of losing touch with religion through the rise of the scientific worldview plus the Romantic notion that, to truly live, you must experience your life as highly meaningful, formed a perfect storm that gave rise to the concept of the existential crisis and conditions endemic to our modern culture today, a society where the lack of meaningfulness can become all-consuming. Carlyle, Kierkegaard, and Schopenhauer may have inherited the yearning for a grand meaning from the Romantics, but the increasingly secularized world they were living in made it seem impossible to attain such meaning.

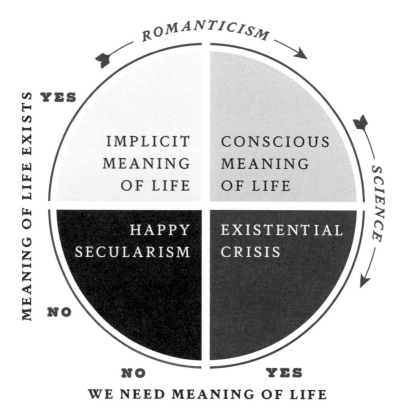

We can better appreciate the impact of romanticism and science by examining these influences in the 2 x 2 matrix depicted above. As regards the matrix, humanity has spent most of its history inside the square of implicit meaning of life. To our ancestors, every creature had a role to play in the enchanted world, and life was so obviously meaningful that—save for the occasional philosopher—it wasn't necessary to even contemplate the question of meaning. Then there's the category of people who are aware that life as such has no meaning but who aren't looking for a sense of cosmic meaning anyway; the whole affair doesn't seem to matter. As far as this happy secularism category is concerned, ignorance can be bliss. I know several people who don't believe in a higher meaning or purpose nor feel anything is lacking from their lives because of their disbelief. Many of them have grown up in nonreligious families. Never having much to do with any religion, they don't feel that they're missing something. On the other hand, I also know several people who feel that life ought to have a higher purpose and for whom God or their faith provide that higher purpose, and this group, in the conscious meaning of life section of the matrix, also don't feel anything is lacking from their lives because they have their faith. The fourth category, existential crisis, is the most tragic as it is composed of those who yearn for purpose while fearing that such a purpose might not exist. The individuals of this group are most susceptible to facing the absurd and thus falling into an existential crisis, or, if history has proven anything, of becoming poets or philosophers. The scientific worldview may have destroyed the enchanted world for them but Romanticism and the religious legacy have led them to believe that such meaningfulness is needed. Unfortunately, this is the category too many of us belong to today—or at least we've dipped into it during our darker moments.

The tension between the grand Romantic promise of a higher purpose and a secular Universe unable to make good on that promise has transformed the question of meaning into the Holy Grail: its pursuit is noble, and while everyone wants to find the answer, we've resigned ourselves to the probable fact that none exists. We cope with the discrepancy of having a question that absolutely needs an answer but clearly doesn't have one by turning the whole question into a joke. In the closing scene from the film *Monty Python's The Meaning of Life,* actor Michael Palin unceremoniously accepts a gold envelope, the contents of which contain the answer to life's biggest question. He reads, "Well, it's nothing very special. Try to be nice to people, avoid eating fat, read a good book every now and then, get some walking in, and try to live together in peace and harmony with people of all creeds and nations." Author Douglas Adams takes it one step further in *The Hitchhiker's Guide to the Galaxy,* in which a supercomputer expressly built to compute the answer to the grand question about life and its meaning spits out its meaningless solution: "forty-two."[94] Both answers only point to the ridiculousness of the question itself—we expect a kind of all-clarifying answer we know doesn't even exist.

The Revolutions of Modern Humans

Besides the rise of Romanticism and the scientific worldview, several other revolutions in the Western world helped influence and shape the way modern man understood meaning and his place in the Universe over the past five centuries. Three influences, in particular, stand out.

First, humanism elevated the role of the self. No longer at the mercy of gods, spirits, or fortune, man was, instead, self-reliant and could navigate his way through life unfettered by such external forces. This transition is most visible in René Descartes's *Meditations on First Philosophy,* published in 1641, in which Descartes used radical doubt to demonstrate the existence of God and the immortality of the soul beyond reasonable doubt.[95] His intentions and conclusions were thus religious—putting belief in God on indubitable foundations. However, his method harbored a time bomb. By proving the existence of God *through reason,* he put the thinking human before God. The tables were turned: human reason became the foundation of God's existence and not vice versa. What Descartes and his contemporaries didn't realize, however, was that if reason could be enlisted to prove God's existence, it could also be used to disprove it.

Second, individualization redefined the relationship between the individual and society. In the premodern world, the group came before the individual.[96] To the extent that there was an individual, he was defined through roles—his position within the family, his social class, his occupation, and so forth—that helped shape the community at large. A person's duty was to fulfill her role-based duties, no matter her individual emotions, dreams, or wishes. In fact, the whole idea of a private, "inner" self beyond one's public self started to appear in the literature only from the sixteenth century onward.[97] The reformation initiated by Martin Luther in 1517 emphasized the direct, unmediated relationship between the individual and God, and the role of one's own conscience. This theological revolution played a major role in making people focus on the individual and one's inner convictions as separate from the group. Luther couldn't have possibly foreseen how his ideas would morph into our modern preoccupation with the individual. Sadly, our obsession with celebrating individual emotions, dreams, and wishes is often a detriment when it comes to good of the group.

Another modern sentiment largely alien to the medieval person is the belief in progress through human effort. Both the scientific worldview and the Industrial Revolution taught people that the world was a far more controllable place than what was previously thought. Instead of being a place where wild cosmic forces could only partially be tamed through rituals and sorcery, and where the order of things was static and stable, the world obeyed certain laws of natures. As such, if man understood the science behind these laws, perhaps he could make them work in his favor and advance his own progress and projects. The industrious modern man came to understand the world as being something to be mastered and controlled—and soon the various inventions and real improvements to life conditions brought on by the Industrial Revolution made the idea of progress feel like a natural course of life.[98]

These three transformations—humanism, individualism, and humans' sense of mastery and progress—worked in tandem with the scientific worldview to create a completely new understanding of humans' place in the Universe. Of course, myriad other transformations were also simultaneously taking place: urbanization, the rise of the mercantile class, and industrialization that uprooted people from their land and communities.[99] The importance of the new democratic political systems that the US and France initiated in the later eighteenth century shouldn't be under-estimated: the legitimacy of the ruler no longer came from God but from people, the citizens. The cumulative result of these various revolutions in thinking was a brand-new worldview that left people increasingly on their own regarding life values and a sense of purpose and meaning. The individual separated herself from the group, followed her inner convictions and desires, and believed herself to be both capable and responsible in achieving her own progress guided by self-selected values. In short: We lost touch with our traditional communities and with the idea of enchantment—even God. We entered the age of self-reliance.

THE AGE OF SELF-RELIANCE

"Belief in God isn't quite the same thing in 1500 and 2000 . . . there are no more naïve theists, just as there are no naïve atheists."

—CHARLES TAYLOR,
A Secular Age, 2007

The US lags behind many European countries in terms of loss of faith, but even here things have been accelerating. Currently approximately fifty-six million US citizens live outside any formal religion[100] but if the present trend in secularization continues at the same pace, the majority of Americans will be unaffiliated with any religion by 2050. This statistic is relatively shocking given that, for a long time, religiosity in the US was rather stable. Starting a bit after 1990, however, there was a sharp growth in the number of unaffiliated, making them by a clear margin the most rapidly growing religious group in the United States. The shift is especially clear in younger generations, with 36 percent of millennials born after 1981 already unaffiliated.[101]

In several European countries, not believing in God—and being open about it—is today business as usual. The Czech Republic is probably the most atheistic country in the world, with 40 percent of the population firmly denying the existence of God, with Estonia coming in second.[102] In France, Germany, and Sweden, more people are convinced about God's nonexistence than existence.[103] And although the majority of people in many countries such as the US still believe in God, it's become more common and acceptable to change one's faith—a leap that would have been unfathomable to the medieval person and perhaps, even, to your grandparents' generation. According to recent Pew research, 42 percent of Americans currently have a different religious affiliation than they did in childhood. Social scientists Robert Putnam and David Campbell note in their study on religiosity in the modern-day US that "it seems perfectly natural to refer to one's religion as a 'preference' instead of as a fixed character."[104]

Even among those Americans and Europeans who maintain a firm belief in God, the notion of "belief" has, in and of itself, become much more modern in what I would argue are four specific ways. First, belief has become conscious, as philosopher Charles Taylor, an expert of the changing nature of religiosity, has emphasized.[105] Today, instead of being so self-evident that one is not even aware of alternatives, belief has become a conscious choice; it's a public declaration of one's faith made with an awareness of the alternatives.

Second, by and large, both believers and nonbelievers have largely accepted the same natural explanations for how the world works. A firm believer may see divine intervention here and there, but most everyday occurrences today are explained by electricity and other natural- or human-generated forces rather than by spirits. When a believer's car is broken, he or she looks for a mechanical malfunction rather than spiritual mischief.

Third, we live in a religiously diverse world. Whether or not you've gone through a religious change yourself, you likely know someone who has. In your workplace and in your community, you most likely encounter people of varying religious backgrounds. In the 1950s, it was still possible to live your life in a way where your spouse, your work colleagues, and your neighbors attended the same church as you. Since then, however, the world has diversified, and especially in urban settings it's become so normal to interact with people from various religious backgrounds that we don't even notice it anymore.[106]

Fourth, instead of relying on God and various spirits to achieve success, we've become increasingly self-reliant on our human capacity and ingenuity to overcome challenges. We may offer our thoughts and prayers, but we know they're not enough on their own to address large-scale social problems. We might ask a priest to bless a new airplane, but we still want to be sure it was designed by a competent engineer and checked by the proper authorities aware of safety issues. In private discussions about how to solve problems in our own lives, and in public discussions about how to solve big political issues—climate change, health care, political division—the solutions are justified by evidence and reason not divine revelation and consultations with spirits.

This is the age of human self-reliance, and it's what separates the modern human condition from most other historical periods.

Today, we have the freedom to choose whether or not to believe in God, and what kind of God or spirituality to believe in. Our worldview is now a matter of personal choice, one we can tailor-make to suit our individual needs, biases, and beliefs.

ONLY A TORCH FOR BURNING?

> "Everything must have a purpose?"
> asked God.
> "Certainly," said man.
> "Then I leave it to you to think of one
> for all this," said God.
> And He went away.
>
> —KURT VONNEGUT, *Cat's Cradle*, 1963

So here we are, a century and a half after Carlyle, Tolstoy, and the original Romantics. . . . Are we any closer to having the answer to the question of the meaning of life they so desperately sought?

I'm afraid the answer's no. While the triumph of the scientific worldview in our ability to control the environment, to fight diseases, and to produce life-enhancing machines and devices is completely unprecedented in the history of humankind, there's still the same gaping hole in the middle of it as existed two centuries ago: the scientific worldview seems to leave us empty-handed as regards human values and how to assign some meaning to our lives. The trouble with science was best expressed by Carlyle, who wrote, "Only a torch for burning, no hammer for building?"[107]

The trouble is, when left to our own devices, we aren't sure what to value or pursue in life. Inventing our own values from scratch turned out to be a harder task than Nietzsche, Sartre, and other existentialists believed it would. They thought that, finally freed from the shackles of tradition, the self-governing, so-called übermensch—the individual above the petty morality of the ordinary people—would create one's own values. More typically, however, people who lack a value framework search for direction and guidance from whoever is still willing to give it. A social position previously upheld by priests, elders of the tribe, and community leaders has been taken over by self-help gurus, self-interested politicians, advertisers, false prophets, and the like. The old worldview is gone, but we're not sure if we like, let alone trust, its replacement.

The cultural-historical process from the enchanted certainty of the medieval worldview to a modern disenchanted, human-centric, and doubt-stricken worldview took several centuries. But many individuals go through this mental process during a single lifetime—even within a single period of their lives. As a cultural process, the transformation is an interesting historical development. As a process within an individual, it's often tragic, leading to a crisis that can feel completely overwhelming. Thankfully, though, there's a path out of this mire, and whether you know it or not, you already possess many of the tools needed to both find and create lasting meaning.

THERE
MEANING
WITH OR
MEANING

"What makes people
find a universal meaning
then end up by saying it
empty of meaning. There
meaning for all, there is
we each give to our life,
meaning, an individual
individual novel, a book

IS *IN* LIFE

WITHOUT THE

OF LIFE

despair is that they try to

to the whole of life, and

is absurd, illogical,

is not one big cosmic

only the meaning

an individual

plot, like an

for each person."

—ANAÏS NIN,
The Diary, Vol. 2, 1967

7

The Universe may no longer be enchanted, but humans still yearn for meaning. Is there a way out of this conundrum? Fortunately, yes. And it starts with understanding the difference between the meaning *of* life and the meaning *in* life.

When people ask about the "meaning of life," they're typically looking for some kind of universal meaning, a meaning that applies to life in general. The meaning of life is about a purpose externally imposed upon life, something given to living beings, presumably from a god or the cosmos above. The meaning of life is thus about something beyond that life in question justifying its meaningfulness. To find a satisfying meaning of life, people have traditionally turned to whatever religious tradition they subscribe to—Christians can consult the Bible; Muslims, the Qur'an; Hindus, the Bhagavad Gita, and so on—to find what role they play in God's master plan or what their proper place and purpose is in the cosmos. What's common here is that some authority from above provides the meaning to human life.

Then there's the question of the meaning in life. This is much more personal. Meaning in life is about what makes your life feel meaningful:[108] It's about experiencing meaning within your own life. This question is thus not about any universal values but what values, goals, and purposes you personally find worthwhile that could help guide your life. It's about identifying or creating whatever it is that makes your own life feel worth living.

While it may be true that finding a meaning of life is impossible without religious belief or some other kind of belief in the supernatural,[109] there's nothing that restricts you from experiencing your own life as valuable and meaningful with or without the supernatural. Life is first and foremost something you experience, not something you impartially observe. Thus the ultimate question about life is how to experience it as meaningful, not whether it's meaningful when impartially observed from above. I, for one, am happy to give up the quest to find the meaning of life in order to concentrate on the question that really matters: How do I find meaning in my life?

Unlike the meaning of life, which easily turns into metaphysical quibbles far removed from your everyday existence, the meaning in life is something you attend to every day, through your every action. Whenever you make a choice, consciously or unconsciously, you tend to choose the option that, in some way, feels more valuable to you. Your tiny and grand life choices thus are your answer to what things make your life feel more valuable and meaningful. You often don't have any specific theory to guide you in these choices yet somehow some choices and some experiences feel more meaningful to you than others. Value and meaningfulness are thus already imbued within your life as lived experiences. In fact, life is full of meaningful moments: hugging a good friend after a long separation, cooking a good meal for your family, accomplishing something as a team at your place of work, noting that you've grown more competent in some hobby you're passionate about, helping someone in a moment when they truly need it. To experience these moments as meaningful, you don't need any theory or rational justification. You can simply experience their inherent meaningfulness.

The trouble with much of the philosophy about the meaningfulness in human life is that it takes "from above" as its starting point. That is, it's taken an impartial and removed look at life and then tried to logically deduce some meaning onto it. But by being impartial, this perspective has already lost the meaningfulness that you inescapably experience within your life. Meaningfulness happens within living rather than outside of it; experiencing meaningfulness is as natural as experiencing warmth or compassion. Thus, rather than looking from outside of yourself, you can start your examination of meaningfulness from within and by investigating those experiences of meaningfulness that are already a part of your life.[110]

When you shift your attention to focus on the meaning in life—to those experiences that make your life feel meaningful—you'll soon realize that you have many relationships, experiences, and emotions in your life that already feel meaningful to you regardless of a rational explanation as to why. As regards the existentialists, perhaps it was not the superstar of the movement, Jean-Paul Sartre, who was right about human meaning, but Simone de Beauvoir, whose philosophical contributions are too often neglected. Contrary to Sartre's fantasies about a self-sustaining, self-governing individual totally separate from others creating his own values out of thin air, Beauvoir emphasized how each of us is already embedded in a particular situation from which we operate.[111] She sides with Sartre and other existentialists in announcing that we shouldn't seek the guarantee of externally imposed values. If life is to be valuable, it's not because someone out there made it valuable, but because you yourself experience it as such. But instead of inventing values out of thin air, her more situated existentialism emphasizes the fact that we already value and desire certain things.

For her, existentialism in its extreme forms "encloses man in a sterile anguish, in an empty subjectivity" and is incapable of "furnishing him any principle for making choices."[112] Instead, as human beings, we're already situated; we already have many values, convictions, and desires. It's from here, from life as it is experienced, that you can and should start your ethical growth and your quest for better values. To make room for such growth you must remain open to revising your values, goals, and commitments. As Beauvoir notes, "Man must not try to dispel the ambiguity of his being but, on the contrary, accept to realize it."[113] Instead of starting from an impartial point of view, or from scratch, you're better off starting from what you currently experience as being valuable and worth committing to—and building on that. At the heart of what Beauvoir calls an ethics of ambiguity is a certain kind of humbleness regarding one's values combined with an openness to learn and grow—and this I see as where the path to a more meaningful way of living starts.[114]

When thinking about meaningfulness, don't start with the big metaphysical questions about the origin of the Universe and so forth. Instead, start from your living experience. Start from where you are right now. Reflect for a while about recent experiences you've had. Think about which ones have been more meaningful than others. Then consider which ones have been less meaningful. Once you've identified the most meaningful experiences in your current life, you can begin to start thinking about how to make life choices that guarantee more of those experiences in the future. If spending time together with a certain person is the most meaningful moment you can think of, how can you be with that person more often? If certain work tasks are more meaningful to you than others, what can you do to build a career path that better utilizes that skill set? Use your own life experience as the starting point to deepen your sense of meaningfulness and fulfillment. And if you're still a little stuck, don't worry. As will be evident in the next chapters, there are a few common core values that can help you identify where most of us typically source meaning in our lives.

Ironically, when we focus on the meaning of life, we tend to separate from one another, drawing sweeping judgments and distinctions that further isolate us in self-referential echo chambers. Your traditional belief system, for example, may grate against mine, and our difference of opinion, if not handled respectfully, can easily create a gulf between us. This isn't sustainable today, not in a globally connected world where we're prone to encounter people from different cultures, backgrounds, and belief systems from our own on a daily basis. If we focus on the meaning in our lives, however, it's striking how much similarity exists in our typical sources of meaningfulness. The same things fill people's lives with meaning all over the world, no matter the cultural context. The human condition truly is universal, and in our global world, it's become important, perhaps now more than ever before, to identify the qualities, characteristics, and needs that unite us. Hopefully, by identifying our basic human condition we can also learn compassion and tolerance to gain a better understanding of one another. Let's take a closer look at how we tend to find meaning in life in much the same way no matter our cultural differences.

MEANING IN THE FACE OF SUFFERING AND DEATH

It is an inescapable fact of being human that, at some point, we will suffer, and some among us will suffer more greatly than others. It's also a sad fact of our humanity that death, eventually, will unite us all—it's a fate none can escape no matter your thoughts on the afterlife, heaven or hell, or indeed, if such planes of existence are real or not. How, then, to make sense of meaning in the face of such grim reality? Does death, as the final game changer, distort the essence of meaning itself? Has all this suffering, this dying, a meaning?

For many of us, questions regarding the meaning of life are connected to one's religious or spiritual beliefs. If we accept the fact, however, that we will each experience suffering and death in spite of our religious or spiritual differences, we can begin to talk about what it means to find meaning in life even while we're staring down a fate or a hopeless situation that can't be changed. Now, what this book can't offer is a direct alleviation of suffering. There are horrible life conditions and tragedies, and we shouldn't fool ourselves that having to go through them wouldn't make one feel agonized. But if one is simultaneously convinced that human life, in principle, doesn't have any value, this makes coping with any difficult situation even more difficult. What could help is if one is able to find some hope of meaningfulness even in this situation. As Viktor Frankl writes in *Man's Search for Meaning*: "For what then matters is to bear witness to the uniquely human potential at its best, which is to transform a personal tragedy into a triumph, to turn one's predicament into a human achievement. When we are no longer able to change a situation—just think of an incurable disease such as inoperable cancer—we are challenged to change ourselves."[115] This kind of change might be hard, but in dire enough situations it might be our only hope.

In addition to suffering, death is the other showstopper for meaningfulness. Some people seem to be convinced that only eternal life guarantees a meaningful existence. If everything vanishes sooner or later, what's the point of struggling? The impermanence of life, as noted in chapter 2, is one of the key pathways through which we come to contend with the absurdity of life. The fact that our lives are undercut by death is often taken as a destructive argument against life having any permanent meaning. And yet, I see no reason why there could be no meaning without eternity.

Imagine you see a small child drowning in a river. Without thinking, you jump into the water and save him. Your heroic act was
highly valuable. You truly made a difference in the boy's life, and
that is meaningful if anything is. The meaningfulness of your act
doesn't diminish from the fact that sooner or later, perhaps in some
hypothetical eighty years, that boy will still die. Even if he dies at
some point in the remote future, your act made possible all those
intermediate years. Because of you, this same boy got a chance to
grow up to become a healthy and confident man with an interesting
career, a loving family, good friendships, and all the other things life
had to offer him. It could be said that your act of heroism made this
life possible and is further manifested in the lives of everyone with
whom he has a relationship—the boy's parents, siblings, friends, and
his future family.

Meaningfulness happens during living, not after it. While
humans are reflective, we truly only experience life in the present
moment. In this sense, the past is simply a compilation of memories
we experience in the present. The future is the projection of hopes
and predictions we make in the present moment. As philosopher
Gregory Pappas puts it, "foresight, hindsight, and present obser-
vation are all done in the present for the present."[116] He quotes
approvingly philosopher John Dewey, who summarizes his recipe
for meaningfulness as follows: "So act as to increase the meaning
of present experience."[117] Meaning in life is about the experience
of meaningfulness. And as far as I know we do the experiencing in
the present moment, right here and now. The meaningfulness you
experienced in 2020 will not be taken away from you in 2030. Life is
composed of temporal moments, some of which are more meaningful than others. And them not lasting forever doesn't detract from
their meaningfulness, as Aristotle pointed out: "Moreover, the good
will not be good to a greater degree by being eternal either, if in fact
whiteness that lasts a long time will not be whiter than that which

lasts only a day."[118] This is true of life more generally. How much meaningfulness you experience in your life isn't determined in some mystical point in the far-off future. It's determined every day, throughout your life. Meaning in life is something we can experience only while we're alive.

Instead of making our lives feel less meaningful, an awareness of death can make our lives feel more meaningful and uniquely valuable. Knowledge that your time on Earth is limited may help you appreciate your days even more. This is why near-death experiences that make people acutely aware of the limitedness of existence—overcoming a mortal illness, for example—often lead people to reprioritize and change their lives in dramatic ways. It's sometimes jarring to realize that your life is a far cry from what you really want it to be: you can't float through life postponing the moment when you stop to contemplate what truly matters in life. If you want to take charge of your own life, it's best to do it before it's too late. No wonder that the ancient Stoic philosophers, twentieth-century existentialist philosophers, and Buddhists have all recommended death awareness as an important life exercise. *Memento mori*—remember, you will die—has been the slogan of various ascetic and spiritual traditions throughout Western history. Life is short. As Samuel Beckett wrote in *Waiting for Godot:* "They give birth astride of a grave, the light gleams an instant, then it's night once more."[119] The best remedy is to resolve to make life choices that ensure that while the light still gleams, your remaining days, weeks, and years are worth it.

But how do you make sure that your remaining days are as meaningful as you want them to be? To answer that question, let's look at the factors that typically make everyday experiences feel meaningful.

BOOKS, DEATH, AND NINETY YEARS
OF HUMAN LIFE IN WEEKS

For me, the awareness of my own mortality and limited time on Earth didn't materialize through a near-death experience but rather, perhaps appropriately enough for a philosopher, through books. Being an avid reader, I often have several piles of books lying about—some of which are in the "must-read" category and others are designated to the "I'll-read-soon" pile. One day, after having moved into a new apartment, I was contemplating what to do with all the books that didn't fit in my new space. I had a sobering thought: there was a very good chance I'd go to my grave without ever having read Baltasar Gracián's *The Art of Worldly Wisdom,* Gunnar Myrdal's *Beyond the Welfare State,* or Marcel Proust's *Contre Sainte-Beuve.* In fact, when I die, many of my books will have gone unread. Here I was, standing in my cellar, staring at the bookshelf, and coming to terms with the limitedness of human existence—all because of a few unopened bestsellers. I don't want to be too dramatic about it, but I must admit that I let my fingers slide across their spines as a gesture of farewell.

You could say that books—or the knowledge that my appreciation of them is finite—is what awakened me to come to terms with an understanding of my own mortality. For my friend, it may have been the realization that there are only a limited number of steaks he'll get to eat from his favorite restaurant: if he visits three times a year and has thirty more years to live, he's got less than a hundred steaks to enjoy. Author Tim Urban, of *Wait But Why* fame, realized that, despite being in his early thirties, he would get to enjoy taking a dip in the ocean—something he does about once a year—surprisingly few times. "As weird as it seems," he writes, "I might only go in the ocean sixty more times."[120] Life is fleeting; it's best to make every moment count. The trick is to savor your limited days by remembering that this one life, as best we know, is all you've got. To help put this idea into perspective, you can count how many weeks you still have left, if you live, for example, to the ripe age of ninety, as Urban has proposed. How many of those weeks will you make meaningful? As the movie character Ferris Bueller famously said, "Life moves pretty fast. If you don't stop and look around every once in a while, you could miss it."[121]

A Ninety-Year Human Life in Weeks

WEEKS OF THE YEAR

AGE

Based on waitbutwhy.com post "Your Life in Weeks."

Building a
Value

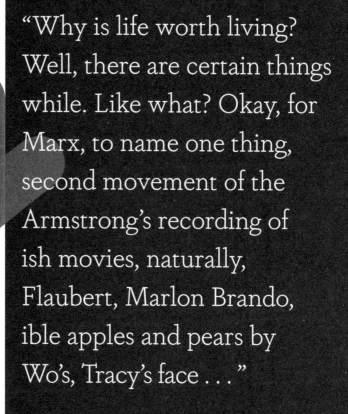

"Why is life worth living?
Well, there are certain things
while. Like what? Okay, for
Marx, to name one thing,
second movement of the
Armstrong's recording of
ish movies, naturally,
Flaubert, Marlon Brando,
ible apples and pears by
Wo's, Tracy's face . . ."

Personal
System

That's a very good question.
I guess that make it worth-
me, I would say, Groucho
and Willie Mays, and the
Jupiter Symphony, and Louis
'Potato Head Blues,' Swed-
Sentimental Education by
Frank Sinatra, those incred-
Cézanne, the crabs at Sam

—WOODY ALLEN,
Manhattan, 1979

Meaningfulness—what you find meaningful and how you want to organize your life around it—is deeply personal and highly subjective; it wholly depends on your experience moving through this world with your unique psychological, genetic, and social background and makeup. People find their lives meaningful in a plethora of ways. What sparks meaning for one person may fall flat for someone else. For a die-hard Packers fan, watching the Sunday game with a few old high school buddies could be the very definition of what makes their life worth living, while for Woody Allen's character in the movie *Manhattan*, it's the second movement of the *Jupiter Symphony* and a few other choices fitting a highbrow New Yorker. Certain areas in the forest where I spent most of my childhood summers are borderline sacred to me but to anyone else, they're just a collection of rock, moss, and a few trees. Denying this richness of possible sources of meaning doesn't serve any purpose; it's best to accept that we each have peculiar sources of meaning that make sense only to ourselves because of our individual history. However, behind this apparent richness, there are discernible patterns in our commonality.[122] As creatures of evolution, humans have been shaped in certain ways, which means that some sources of meaning come easier to us than others and have near-universal appeal. Basing our quest for meaning on these sources might be the path to the most robust meaningfulness available to us fallible human beings.

FROM ANIMAL PREFERENCES TO HUMAN VALUES

"Before life began, nothing was valuable. But then life arose and began to value— not because it was recognizing anything, but because creatures who valued (certain things in particular) tended to survive."

—SHARON STREET,
A Darwinian Dilemma for
Realist Theories of Value, 2006

From the simplest creature to the most biologically complex, animals prefer certain experiences over others: pleasure before pain; satiety over hunger. Different species have unique preference systems based on those experiences that have proven essential for their survival. An antelope is drawn to the watering hole, but upon sighting a lion immediately runs away. Humans are no exception. Our very humanity already has several built-in preferences that guide our behavior to help ensure we get what we need to survive. We share many of these basic instincts with other animals. To better understand what sets us apart, however, we can look to the work of philosopher John Dewey.

Considered a father of functional psychology and one of America's most important public intellectuals during the first decades of the twentieth century, Dewey made an influential distinction between "prizing" and "appraising."[123] Prizing, the act of preferring one thing over another, is primarily emotional and happens naturally. Every animal engages in some form of prizing. Humans, however, don't settle for mere preferences. We also engage in appraising, whereby we consciously evaluate and examine our preferences, and make judgments as to which values are, in fact, meaningful to us and worth valuing. We want more from our values than their existence; we want them to be justified. It is only after a value has withstood a rigorous test of reflective scrutiny that it becomes something we endorse, either individually or socially, and use as a metric to measure the rightfulness of our actions and behaviors. Animals have instinctual and learned preferences, what we may call proto-values. Only humans have proper values, understood as something one is consciously committed to and reflectively endorses. Values thus don't exist outside of human experience but within it, as the general preferences we willingly hold in highest regard.

Human values are like trusty tools.[124] We use them every day to help guide our personal decisions and to better adhere to social and moral codes. But our commitment to our most sacred values is typically so strong that we're willing to base our lives on them—in extreme cases even die for them. They are the backbone of a meaningful existence. A person without values is like an animal, following one's instincts without any constraints. Values elevate our existence above the animal world and beyond its instinctual game of survival. As the movie character Solomon Northup memorably states in *12 Years a Slave:* "I don't want to survive. I want to live"— mere survival is not enough to make a life worth living. It is by both recognizing and cultivating values worth committing to that we're able to imbue our lives with a strong sense of meaningfulness.

But how do we—how do you, right now—identify the values that give life its specific meaning? In the past, cultures have typically prescribed value systems to which everybody is expected to adhere. Today, however, you're surprisingly free to choose your own values. How do you make correct use of this freedom? How do you strengthen your values into a ringing endorsement of all that you are?

SELF-DETERMINATION THEORY— FROM BASIC PHYSIOLOGICAL NEEDS TO BASIC PSYCHOLOGICAL NEEDS

Self-determination theory, cocreated by professors Edward Deci and Richard Ryan, has grown into one of the most empirically researched theories of human motivations and needs, backed up by hundreds of studies. It is based on the simple idea that humans are inherently curious, self-motivated, and growth-oriented beings who not only respond reactively to environmental stimuli but also actively self-regulate their lives toward growth and integrity following their internal motives, goals, and values.[125] In a word, humans are active. When we orient ourselves in the world, we don't just look to satisfy our physical needs—food, water, shelter. We naturally and intuitively look for more in life: a chance to express ourselves, to grow and utilize our skills, to feel connected to others, among others. Humans don't settle for mere survival. As active creatures, we naturally set tasks and goals for ourselves. We search out challenges for ourselves and create innovative solutions to overcome our limitations and all the while, we're guided by what Deci and Ryan define as our basic psychological needs, that is, the "innate psychological nutriments that are essential for ongoing psychological growth, integrity, and well-being."[126] As an acorn needs soil, sun, and water to grow into a healthy oak tree, human beings need certain experiences to develop into psychologically healthy individuals.

Research on basic psychological needs aims to identify those experiences humans require to feel good and to flourish in life. If your basic psychological needs are momentarily frustrated, you feel bad and anxious. However, when you're able to satisfy your needs, it not only feels good but it also contributes to your overall psychological integrity and well-being. At the same time, these needs help you engage in activities that encourage growth in your life. They're growth-oriented in the sense that they aren't immediately connected to your survival the way your physical needs are. Instead, engaging in them builds skills, resources, and social connections that may feel valuable in and of themselves, but often also come to use when facing challenges later in life. Luckily, it's not only fulfilling to answer your growth-oriented needs, but it's also something most of us enjoy doing. And here's the kicker: building your core values upon these growth-oriented human needs is the best recipe for a highly meaningful life.

THE BASIC PSYCHOLOGICAL NEEDS UPON WHICH TO BUILD YOUR VALUES

Self-determination theory recognizes three basic psychological needs: autonomy, competence, and relatedness.[127] When these three needs are satisfied, people experience more well-being and intrinsic motivation—and, indeed, more meaning in life.

Autonomy is about being the author of your own life. You're able to make volitional choices to live according to your own preferences, engage in activities you find personally interesting and that express who you are, and pursue goals you find worthy. Competence is about having a sense of mastery in your life. You're confident in your abilities, feel skillful in the activities you undertake, and trust that you'll achieve your goals. Competence doesn't have to be static: you can experience the feeling of competence as you're learning something new or further honing your skills. Relatedness refers to a sense of being connected to others, caring for them, and also feeling cared for. These three needs form the triad of psychological needs upon which self-determination, well-being, and one's sense of meaning can unfold. Yet, as regards meaningful living, I feel that there's an important component missing from the equation.

"Only by knowing the kinds of beings that we actually are, with the complex mental and emotional architecture that we happen to possess, can anyone even begin to ask about what would count as a meaningful life."

—JONATHAN HAIDT,
Happiness Hypothesis, 2006

Benevolence is commonly defined as "a disposition to do good." It's about a desire to have a positive impact on the lives of other people, society, or the world at large.[128] It can be as mundane as making a fellow neighbor smile or as life-changing as rescuing someone from a burning building. Big or small, when you feel your life and your actions make a positive difference in the world, you feel you matter—your contribution is meaningful. While research by myself and others hasn't confirmed benevolence as a psychological need per se,[129] it has been shown to be an important contributor to our sense of well-being and especially to our sense of meaningfulness.

Accordingly, as regards what makes life meaningful, I'm willing to bet my money on autonomy, competence, relatedness, and benevolence.[130]

As a human being, you thus have certain innate psychological needs. When you're able to satisfy these needs, you gain a deep, almost visceral, sense of fulfillment and satisfaction. The next step toward experiencing more meaning in your life is to build upon this foundation, adopting these needs—and their fulfillment—as the core values of your inner life compass.[131] Every innate psychological need should thus be accompanied by a corresponding reflectively endorsed value, and pursuing these values is a path toward a sense of meaning in life. Thus the need for autonomy is accompanied by valuing authenticity and self-expression; the need for competence by valuing mastery and excellence; the need for relatedness by valuing belonging; and the potential need for benevolence by valuing contribution.

In seeking guidance on how to live a meaningful life, my suggestion is to go with this quartet. They have strong intuitive appeal, and endorsing them leads to good outcomes, individually and socially. Their embeddedness in human nature makes them robust and self-justificatory enough to provide worthy answers to the question of meaning. It also means that they're endorsed across cultural, religious, economic, and other boundaries.[132] In valuing what all humans inherently already find meaningful, not only can you recruit those around you to support your value system but you also have a far better chance of finding common ground even among people from different backgrounds. No matter where you come from, what religion you subscribe to, or any other particularity or difference between yourself and your neighbor, our basic human nature unites us, and from it you can source the strongest meaning and value in life. Besides being intuitive and worth committing to, these values are also easy to put into practice.

You Get What You Pursue

"Be careful what you wish for,
you just might get it."

—PROVERB

People can chase many goals and values, but research has shown that pursuing goals aligned with our basic needs is beneficial for our sense of well-being and meaningfulness while pursuing goals not aligned with them can, in some situations, lead to an increase in ill-being. For example, one study conducted by Richard Ryan, Edward Deci, and Christopher Niemiec at the University of Rochester a few years before I arrived there asked graduating students to think about pursuing certain goals in life.[133] Some goals were well aligned with growth-oriented basic needs: the desire to have good relationships with others, to contribute to the community, and to develop more as a person. Other goals were of a more extrinsic nature like the pursuit of wealth, fame, or good looks. A year later, these students were contacted and asked how their current status measured up to their goals. Looking at the results, the researchers observed, first, that you get what you pursue. Seeing certain goals as important tended to lead to more progress in the achievement of those goals. People who valued maintaining good relationships felt that their relationships had deepened. People who valued good looks felt that their looks had improved. Nothing surprising here: if you value something, you tend to put in effort to get it, and if you put in effort, you also tend to make progress toward that goal. But the researchers also looked at the well-being effects of goal pursuit. It turned out that making progress in goals aligned with the basic needs increased people's sense of well-being. However, progress in achieving extrinsic goals not related to these needs didn't increase well-being—in fact, it even slightly increased feelings such as anxiety and other negative emotions. So even though students seeking wealth, fame, and good looks had made progress in achieving goals important to them, this progress contributed to them feeling ill rather than improving their sense of well-being. Choose your goals wisely—those aligned with the basic needs of autonomy, competence, relatedness, and, in this case, benevolence, can improve your well-being. Goals not aligned with these basic needs may make you feel worse—even when you achieve them.

Part Three:

PATHS TO A MORE

Meaningful

INVEST IN

REL

TIO

SHI

YOUR

A-N-P-S

"One's life has value so long as one attributes value to the life of others, by means of love, friendship, indignation, and compassion."

—SIMONE DE BEAUVOIR,
The Coming of Age, 1970

A philosopher walks into a bar, and a barfly asks him, "What's the meaning of life?" That philosopher is me, and it's an inevitable question once people find out what I do. I've had this happen enough times to have a one-liner ready. I'll first explain that it is not about the meaning of life but the meaning in life, before delivering the punch line. It has two parts, the first of which is the following: meaning in life is about making yourself meaningful to other people. It's that simple. Forget the meaning of life. Your life becomes meaningful to you when you're meaningful to other people: by helping a friend, for example, by sharing a special moment with someone you love, or, more simply, by connecting with a well-intentioned philosopher through buying him a much-needed beer.

When we sense that our lives are meaningful to other people, we're able to see the value in our own lives. The Universe may be silent, but our friends and family, our colleagues and community fill our lives with their voices, energy, and vitality. And the people to whom we are most meaningful are those who care most about us. As philosopher Antti Kauppinen has argued, for those who love us, we are irreplaceable: even though anyone can buy a present for a particular child, "it will not have the same significance as a handmade gift from a parent," as he writes.[134] In close relationships, we play a unique and irreplaceable role for the other person often simply by being there.

If we know anything about human nature, it's that we're social animals. In "The Need to Belong," an influential review article published in *Psychological Bulletin* in 1995, Professors Roy Baumeister and Mark Leary made a claim that has since become a broadly accepted—and seemingly obvious—thesis in psychology: "A need to belong is a fundamental human motivation."[135] We evolved to live in groups and to care for each other; the instinct to build strong social relationships lies deep within our humanity.

Our social nature, however, goes deeper than merely caring about others: it's in our nature to have, as the locus of one's life, not *me* but *we*. Being in a close relationship has been described by psychologists as a state of "including other in the self."[136] Indeed, neurological research has demonstrated that thinking about oneself and thinking about a loved one activate certain regions in the brain that aren't activated when thinking about a stranger.[137] The brain is wired to be social, and humans are designed to live together with others. As the French philosopher Maurice Merleau-Ponty has beautifully explained: "We are collaborators for each other in consummate reciprocity. Our perspectives merge into each other, and we coexist through a common world."[138] Although our Western, individualistic culture has habituated us to carve especially clear boundaries between the self and other, being able to be so separate from others is a cultural achievement rather than our typical way of being. We care about the well-being of those close to us almost as much as we care about our own well-being. Sometimes, as in the case of being a parent, we may care about a child's well-being more than our own. No matter what scientific field we turn our gaze toward—biology, neurological research, evolutionary research, social psychology, behavioral economics, even primate research— we find evidence of our need to form close and caring relationships with others, and how in these relationships the boundary between the self and the other starts to loosen.

Ample evidence shows that relatedness is, indeed, a key source of meaning for us. When researcher Nathaniel Lambert from Florida State University asked a group of undergraduate students to, in his words, "pick the one thing that makes life most meaningful for you," two-thirds of the respondents either named a particular family member or cited, more generally, their family.[139] As a category, "friends" came in second as the most frequently mentioned source of meaning. Pew Research Center got similar results when four thousand Americans were asked to describe in their own words what provides them with a sense of meaning: 69 percent mentioned family and 19 percent mentioned friends.[140] Other research has similarly shown that feeling close to one's family and friends is associated with an enhanced sense of meaning in life, and thinking about people "with whom you feel that you really belong" leads to higher ratings of meaningfulness.[141] Family, friends, and other close relationships are, for many people, key sources of meaningfulness in their lives. The opposite is also true: being socially excluded leads to feelings of meaninglessness. For example, researcher Tyler Stillman and his colleagues recruited a group of students to participate in a study allegedly on first impressions. The 108 students self-recorded a few minutes of video introducing themselves.[142] The researchers then supposedly showed the videos to other students and asked whether or not anyone wanted to meet the video makers: no one wanted to meet them. (In actuality, no one watched the videos; the researchers simply told the video makers they were rejected.) The results of the study aren't surprising: the video makers rated their lives as having less meaning than another group that was spared this experience of social exclusion.

But we don't need research to tell us that encounters with other people are a key source of meaning. As a father of three small children (two, five, and seven years old at this writing), I don't have to look far to see which moments in my everyday life are most meaningful—coming home after work, taking the smallest child in my lap, engaging in some rough-and-tumble wrestling with the five-year-old, and holding surprisingly interesting if not intelligent conversations with the seven-year-old. Moments like these are intimate, caring, and full of warmth—and indeed are highly meaningful. So, too, are the private moments I share with my partner, when no kids demand our attention, and we can look each other in the eye and be reminded that—yes—this is the person I fell in love with all those years ago. At the risk of sounding sentimental, the list goes on—old friends, colleagues, my parents, siblings, extended family— as I'm sure yours does, too.

In the modern world, there are luckily also myriad options for people to have strong relationships and connections to one another without necessarily having the proximity of "family." A group of my friends, for example, who have decided not to have children instead live in a collective with other like-minded individuals. A few guys from my soccer team, in turn, felt so committed to this sports community of ours that they recently got tattoos of our team logo. Some colleagues of mine devote themselves to neighborhood activity, volunteering their time, passion, and resources to make their neighborhood more active and community-centric. The beauty of the modern age is that we have the freedom to choose which sources of meaning connect the most to our lives. Unfortunately, as with much of modernity, this is both a blessing and a curse.

ARE WE EXPERIENCING AN EROSION OF COMMUNITY IN MODERN WESTERN COUNTRIES?

> "No one can live happily who
> has regard to himself alone and transforms
> everything into a question of his own
> utility; you must live for your neighbor,
> if you would live for yourself."

— SENECA, *Letters*, circa 65 CE

I got a small glimpse of a lifestyle forgotten in our hectic and urbanized modern world when I spent a week in a small village of two thousand people accessible only by boat on the east coast of Nicaragua. The sense of community and the slower pace of life were immediately visible. I befriended a local man on my first evening and walked around the village with him; it seemed that every fourth person we met was his cousin. We always stopped to chat because no one seemed to be in a hurry. For him, this small village represented the whole of life: he had been born here; he had known these people his whole life; and he would probably grow old and die here, too, buried in the same graveyard as his parents and grandparents before him. The more time I spent in the village, the more I felt this to be the natural way of living instead of the hectic, urban, isolated, and project-oriented lifestyle back home.

Granted, the temptation here is to cast this coastal living as paradise. Being a casual observer and an outsider, I couldn't accurately see the daily dramas or the interpersonal trials and obstacles that were surely present. Getting ill in the village, for example, could quickly turn to tragedy absent the medical care facilities we've become accustomed to in the West. Still, I couldn't help but envy and marvel at their strong social bonds. The villagers were constantly surrounded by people they had known for years; their families and best friends were all within walking distance from one another, and nearly every face they encountered during the day was familiar.

For most of history, humans lived in a way that more closely resembled these villagers than today's Western citizens. Hunter-gatherer tribes were tight and intimate communities. In agricultural societies people tended to stay put, typically living in the same community from cradle to grave. Comparatively, today's Westerners are uprooted and isolated. The extended family has given way to the nuclear family with relatives often living thousands of miles away. Our "closest" relatives are literally no longer very close to us.

The story of community and modernization, however, is not only one of decline. In fact, individualism has given rise to new forms of community previously unavailable to the farmer or the hunter-gatherer. While we might have lost out on the rootedness and the proximity that once characterized communities, we've gained the freedom and the ability to join communities based on our personal values and interests. Being born into a community where one doesn't fit in, for one reason or another, might have led to a lifetime tragedy. Nowadays it often gets better—one can join various communities that better match one's own worldview and interests. A new high school, college, job, or neighborhood often present opportunities for one to build an identity in the eyes of others anew.

Traditional communities have also often been quite oppressive, enforcing certain norms and worldviews, and involving rigid hierarchies where, for example, women have had an inferior status. Although some researchers have sounded the alarm bell around the erosion of community in the US and the Western world, perhaps most famously Professor Robert Putnam with his influential book *Bowling Alone*, the research community seems to be divided as to whether or not any sharp decline in the sense of community has taken place in the last few decades.[143] In fact, some research even suggests that more individualism can be associated with more social capital—the more individualistic a state is in the US, the more prone its people are to trust strangers, belong to various groups, and have higher levels of social capital. And the same holds true at a cross-national level: a comparison of forty-two countries similarly showed that higher levels of individualism were related to more group memberships and a higher trust in strangers. Thus some researchers like Jüri Allik and Anu Realo argue that "individualism is a precondition for the growth of social capital—voluntary cooperation and partnership between individuals are only possible when people have autonomy, self-control, and a mature sense of responsibility."[144]

The relation of modernization and individualism to our sense of community and belonging is complex. Some forms of community might be declining while other forms seem to be increasing. We might have lost the lifelong proximal communities of our ancestors, but we've gained the chance to voluntarily join communities where our individuality is able to bloom with like-minded people. Nevertheless, if we are to make our lives more meaningful—and the lives of our children and grandchildren—we need to work together to strengthen the forms of community available to us. Meaning is about connecting.

Often the best and easiest way to improve your own sense of well-being and meaningfulness is to switch your lens: concentrate less on yourself and more on being connected with others.

ONE FORMULA FOR A WELL-LIVED LIFE

A few years before Sebastian Vettel became the youngest Formula 1 world champion—and a subsequent four-time champion, global icon, and multimillionaire—his doctor, Aki Hintsa, gave him a piece of paper and an envelope. The task: Write down the names of the most important people in your life and why they're important. Vettel did as he was asked and sealed the paper in an envelope. Hintsa advised him to hold on to it, saying, "When success comes, many more people are going to want to be a part of your life. . . . Check this letter to see who your true friends are and remember to stay in touch with them."[145]

Doctor Hintsa used this exercise with many of his clients, often asking them to make a list of people they would take on a several months' long sailing trip or to a remote island. Think about it yourself. Who would you bring? Can you identify the people who are truly important to you and with whom the mere fact of being together is a source of vitality and meaning? Once you've identified them, think about how much time and energy you currently devote to them. Furthermore, think of your interactions with them: Have you been authentic and true with them, with yourself?

Hintsa's clients included many high-functioning, hardworking successful individuals, who, it turned out, often sacrificed and neglected meaningful family relations and friendships for their careers. One business executive, for example, had the habit of taking his wife and children on luxury vacations to exotic locations. There, he'd sign the kids up for various adventures and send his wife to the spa. With his family cleared out, he'd log long hours working outside the office. If that happens every once in a while it's not a big deal, but if this behavior turns habitual, which it had for him, it becomes problematic in terms of the big picture of his life. Children want adventures and spouses may appreciate a good spa day, but if gaining these luxuries means losing out on familial closeness, no amount of exotic vacations can remedy a strained parental or marital relationship. Hintsa's typical advice for both his athlete and executive clients was the same: spending time with the people you love should be at the top of your priority list.

When Finnish researcher Leena Valkonen interviewed eleven- to thirteen-year-old kids for her dissertation about what they wished from their parents, one of the most frequently mentioned wants was time. A twelve-year-old boy wrote, "Parents should remember that family comes first and work only after that." And this family time need not be extraordinary. Most children want to do regular, everyday things with their parents— cooking, talking, cleaning, listening to music, playing catch, just hanging out. As one child put it, "Parents should 'just be' at home."[146]

Back to young Sebastian Vettel. Vettel took the

insight from Hintsa's exercise to heart. A few years later, when he became the global icon he is, everyone wanted a piece of him. He remembered the contents of the envelope and, over the years, carefully protected his inner circle, finding time for his closest family and friends in the midst of frenzied media attention and fame. He's still together with his childhood friend Hanna Prater, with whom he has two kids. He's realized that no matter the level of success you achieve, the secret to a meaningful and well-lived life is having a few good people in your life you can truly trust, care, and love. Whether you're on a cramped sailboat together for many months or shored up on a remote island together or simply withstanding the daily grind of what it means to be human, these are the people you want in your corner, no matter what. They make your life better and you'd do the same for them. If you're lucky, *your* name is written down in an envelope that belongs to someone you love.

10

HELP

OTHERS,

HELP

YOURSELF

"How selfish soever man may be supposed, there are evidently some principles in his nature, which interest him in the fortune of others, and render their happiness necessary to him, though he derives nothing from it except the pleasure of seeing it."

—ADAM SMITH, *The Theory of Moral Sentiments*, 1759

It's Christmas Eve, 1945, in the small town of Bedford Falls, New York, and George Bailey is standing on a bridge, staring into the dark waters below, ready to commit suicide. His business is failing, and his dreams are unfulfilled. He's drunk and desperate and sees no other way out.

Frank Capra's classic *It's a Wonderful Life* is one of the most beloved films of all time. Just as George Bailey is about to jump from the bridge, an angel named Clarence Odbody appears, leaps into the water, and George is obliged to save him, forgetting his own plan in the action. When George afterwards tells him, "If it hadn't been for me, everybody would be a lot better off," Clarence swings into action to show George what life would have been like without his steady and compassionate presence. George then enters an alternative reality that horrifies him: his brother has died, his uncle is institutionalized, his wife is alone, and the whole town is in terrible condition. George quickly understands the positive impact he's made on the lives of his fellow townspeople, including the members of his own family.

Because of him and the many personal choices and sacrifices he's made, many people around him are much better off. His desire for suicide is replaced by a wish to return home and to be there for the people he cares for and who care for him.

Meaning in life is about making yourself meaningful to other people, as previously noted, but there are at least two ways to accomplish that. Not only are we meaningful to those people with whom we have close relationships, but we're also meaningful to those people whose lives we're able to impact in a positive way. This is a crucial second source of meaning in life. When you're able to contribute to the world, even in tiny ways, this tends to increase the sense of meaningfulness in your own life.

Think about a person who has led a particularly meaningful life: names like Martin Luther King Jr., Mahatma Gandhi, Nelson Mandela, or Mother Teresa probably come to mind. What unites them is the fact that their actions have made a positive impact in the lives of whole generations of people. It's their extraordinary contributions to the world at large—often at significant personal sacrifice—that makes us see them as paragons of meaningful lives. Take Nelson Mandela, who, having spent twenty-seven years of his life in prison, stepped up to become the leader of the ANC party in South Africa after a century of apartheid and the oppression of black people. With his policy of forgiveness and nonviolence, he was able to prevent the country from sliding into a civil war that surely would have meant the deaths and profound suffering of tens of thousands of people. Mandela's impact on world history was tremendously positive and is one of the reasons his life is so often held up as the paradigm of a truly meaningful existence. Similarly, when we think about particularly meaningful occupations, professions like firefighters, nurses, and doctors usually lead the pack. Again, what unites these occupations is the fact that they each make a clear positive contribution to the lives of others. Often, when we talk about meaningfulness or the lack of meaningfulness of something, we're actually talking about whether or not that thing has made a positive impact on other people and the wider world.[147]

Accordingly, it's not surprising that when researchers ask people to engage in acts that benefit other people, people tend to report these acts as being meaningful. A few years ago, together with Professor Richard Ryan, I decided to test this theory. [148] We invited students from the University of Rochester to play a simple computer game: The students read one word at the top of a computer screen and four words at the bottom. They were then asked to choose the word from the four that was synonymous with the word at the top of the screen. After playing for twenty minutes, we asked the students to rate how much meaningfulness they experienced while playing the game. If you've ever done redundant tasks at your job, you can probably guess how this activity ranked on a scale of meaningfulness.

Half of the participants played the game as directed. The other half, however, were told that for every correct answer, a small donation would be made to the United Nations World Food Program to help starving people all around the world. Exactly the same game, but with an opportunity to make a positive contribution. Afterward, there was a clear difference between the two groups, with the contribution group reporting the game as being significantly more meaningful than the control group. It would appear we're willing, and perhaps even eager, to overlook boredom when the most mind-numbing tasks can also be used as agents for good. This and other psychological research[149] thus points to a simple conclusion: to experience meaning in your life, find ways to feel you're making a meaningful contribution to the lives of others.

RESEARCH SHOWS THAT HELPING
OTHERS CAN HAVE A SURPRISING
IMPACT ON YOUR HEALTH

> "Compassion is central to human
> well-being, for those who provide it
> as well as for those who receive it."
>
> —MONICA WORLINE AND JANE DUTTON,
> *Awakening Compassion at Work*, 2017

Helping others can have other tangible benefits for your own life besides increasing your sense of meaningfulness.[150] A research team from the University of British Columbia gave spending money to a group of older participants with high blood pressure.[151] Over three consecutive weeks, every participant was given forty dollars each week. However, half of the participants were instructed to spend the money on themselves; the other half were asked to spend it on someone else: buy a gift for a friend, donate it to a charity, et cetera. Before and after these weeks of spending, the researchers measured the blood pressure of both groups. The blood pressure (both systolic and diastolic) of those participants who had spent the money on others had significantly decreased as compared to the participants who spent the money on themselves. Moreover, the decrease in blood pressure was similar in size to the effect of starting a high-frequency exercise or a healthier diet.

So be careful with your helping: it can have insidious benefits to your health. In the worst case, it can make you live longer! A study of 846 older adults compared receiving social support and giving social support as predictors of mortality over a period of five years. While it would be intuitive to think that receiving such support

would be good for oneself, the results showed that it was actually
giving social support that was more predictive of longevity: those
who provided instrumental support to friends, relatives, and neighbors, and those who provided emotional support to their spouse
were more likely to be alive at the end of the study period compared
to less pro-social participants.[152] These results held true even when
the researchers controlled for various demographic factors such as
physical health, mental health, personality, and marital status.

More than ten studies have also demonstrated that regular
voluntary work predicts longevity.[153] Helping someone else can
even buffer against the negative effect that stress typically has
on mortality: among eight hundred participants around Detroit,
stressful events predicted subsequent mortality among those who
did not provide help to others in the past year, but not among those
who did.[154] Also, being the caregiver of an ailing loved one is often
assumed to be burdening for the caregiver. However, while the
stress and sorrow associated with seeing one's spouse fade away is
clearly a heavy burden, active caregiving may have a positive effect
on the longevity of the caregiver. A national study of over three
thousand elderly married individuals showed that those who spent
at least fourteen hours a week providing active care to their spouse
actually lived longer, when controlling for various demographic
and health variables.[155]

And, as if longevity and better health aren't enough, providing
support for others also tends to make the caregiver happier. Professor Elizabeth Dunn from the University of British Columbia has
shown that when a group of people is given five dollars to spend on
themselves and another group is given five dollars to spend on
others, the latter group reports greater happiness afterward.[156] And
this is true not only in her home country of Canada, but also across
the world—from Uganda to South Africa to India—as her colleague
Lara Aknin has shown.[157] Aknin also conducted the study in a

small-scale, rural, and isolated village on the Pacific island of Vanuatu.[158] Even there, purchasing goods for others led to more positive emotions than purchasing goods for oneself. There thus seems to be something rooted in our very human nature that makes it feel good for us to help one another—and this applies across cultures. Various neurological studies have further corroborated the fact that charitable donations do, indeed, activate the reward centers of the brain.[159]

A dose of good deeds toward others can thus be not only meaningful, but also good medicine for improving one's physical and mental health and well-being.

HOW TO CONTRIBUTE

"Compassion has meaning for us all. It enriches us and ennobles us, even those of us who are neither the care givers nor the recipients, because it holds forth a vision of what a good society can be, provides us with concrete examples of caring that we can emulate, and locates us as members of the diffuse networks of which our society is woven."

—ROBERT WUTHNOW, *Acts of Compassion*, 1991

If it's such a powerful source of well-being, health, and meaning in life, what's the best way to make a contribution? First, it's important to remember that it doesn't have to be a grandiose Mandela-like "save-the-nation" type of contribution. As research has shown, even tiny charitable donations can impact one's sense of meaning-fulness. Most of the meaningful contributions we make are small and mundane, yet still imbue our everyday life with meaningful moments. Think about a moment when you were able to truly delight a person you love or care about. Perhaps you prepared an impromptu candlelit dinner for your spouse or helped a close friend struggling with a personal problem.

In teaching on the subject, I often ask my students to perform three random acts of kindness before the next class. Their "contri-butions" have been as varied as offering a glass of orange juice to the mailman, spending an afternoon with a grandparent, and helping a tourist navigate the twists and turns of the neighborhood streets. As a class, we then talk about our various contributions and how we felt doing them; the stories are always inspiring and the exercise has become a course highlight. Not only is it uplifting to hear the inventive ways these students went out of their way for someone else, but it's also touching, with some students reporting that they experienced a deep bond between themselves and the person (or people) they helped. These small acts of kindness made their day, leading to warm feelings of connectedness and meaningfulness and illustrating, in a very real and direct way, that when we help others, we also help ourselves.

If you're looking for a more profound way to make a contribu-tion in life, work is often your best bet. Those of us who are lucky enough to be in professions where we can devote eight or more hours per day doing something that makes a clear, positive impact already have a significant source of meaningfulness built into our lives. Often, it's about simply reminding oneself of the good impact of one's work. In a study, Professor Amy Wrzesniewski of Yale Uni-versity asked hospital janitors to describe how they saw their job. Some reportedly saw their work as "being simply about cleaning" while others "saw the work and themselves as critical in healing patients,"[160] since they helped maintain the hospital's high hygienic standards. Exactly the same job, but two different ways of looking at it—sometimes it doesn't require anything more than an increased awareness to see the contribution we make at work.

If the work you do doesn't involve a grand mission, you can take delight in the rewards of helping out an individual customer or a fellow colleague. A colleague of mine recently proposed that every Friday we should think of one person we want to thank in our company, and then post our gratitude to them in Slack, our internal communications channel. Now, all the small ways we help each other out throughout the week are visible for everyone to see. It's heartwarming to read these messages of gratitude and has helped to build a stronger sense of contribution and community within the company.

Outside of work, you may find your path to contribution through volunteerism, charity donations, helping your friends or extended family, neighborhood activism, or by supporting political causes or campaigns you feel strongly about. Whatever you do, it doesn't have to be complicated. Opportunities to make meaningful contributions abound, if you only look for them. Take the customers of the Soup Place in Melbourne, Australia, for example, who have taken up the policy of "paying it forward" by regularly chipping in an extra $3.50 to purchase an additional meal for a homeless person. The practice has become so popular that one wall of the restaurant is covered in free meal tickets for the taking.

HOW MUCH OF A GOOD THING IS TOO MUCH?

A minor word of warning: too much of a good thing can be bad. If we solely concentrate on the well-being of others, always putting them first, we can run the risk of ignoring our own needs. There are too many tragic stories of people who sacrifice their own happiness in order to serve their family or some grand global cause. Helping is good, but it should be strategic and self-selected, as Professor Adam Grant, an expert on prosocial giving from Wharton University, has emphasized, writing, "There's a big difference between pleasing people and helping them."[161] Helping whoever happens to ask because you don't dare say no is totally different than strategically choosing to help someone you really want to help. Indeed, a number of experiments have shown that while autonomously motivated helping increases the well-being of the helper, this doesn't hold true when one is coerced or forced to benefit others.[162] By learning to say no, we can concentrate on helping when and where our interests and talents are best put to use, and where we can get the biggest impact for our investment. Try to resist helping all the people all the time. Help only those you want to help, and who would benefit most from your assistance, not those who ask the loudest.

No man is an island. As social beings we humans encompass both the desire to look after ourselves and to look after others.[163] That's why the extremes, only serving oneself or only serving others, are detrimental for our sense of well-being. In both cases, part of our humanity is suffocated. Finding balance is the key. But in our era of individualism and unabashed self-interest, reaching such balance often means a commitment to consciously start looking for the best ways to help those around you.

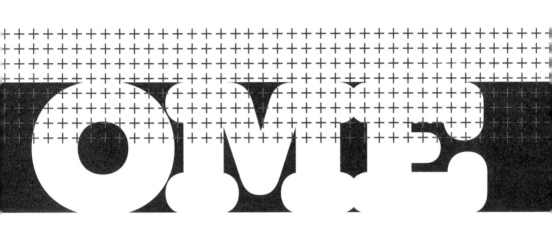

COME

"Trust thyself: every heart vibrates
to that iron string."

—RALPH WALDO EMERSON, *Self-Reliance*, 1841

YOU ARE

My nearly-three-year-old son, like most other toddlers his age, does not walk. He either stays put, solid as a bull, or he runs. Everywhere. There's no middle ground. If he perceives or feels that something is being externally forced upon him, he steadfastly refuses to partake in it. On the flip side, I can tell when something has been lit for him internally—when the force comes from within, as it were. His joy and excitement are palpable, and, in these instances, unabashed running seems to be his only option. With a small child, it's easy to sense whether an activity is externally forced or internally motivated. And we can see even more clearly how the child feels about it: externally forced activity is usually accompanied by a grumpy face and tears; internally motivated activity is usually accompanied by laughter and a smile that can melt hearts.

The greatness of being human is that we have the liberty to do activities that are internally endorsed, something we feel originates from within ourselves. We hold hands with one another. We love. We write and talk. We create.

We sing, dance, laugh, run, climb, and jump. We get excited and immerse ourselves in an activity. Sometimes we are so drawn toward an activity or mode of self-expression that we can hardly wait to begin. We have the capacity to express and realize ourselves, to do what we please. In a word, we can behave authentically. Authenticity is about feeling that one's course of life is self-directed and self-chosen. And this is crucially important for our sense of meaning in life. Meaning is about connection. While the previous two chapters emphasized the importance of connecting with others, it's equally important to connect with ourselves. Otherwise we're just empty shells.

There's an important source of meaningfulness hidden in being able to live in a self-chosen way, doing activities one has self-selected. Philosopher Richard Taylor talks about a "strange meaningfulness," which he uses to describe being able to do activities where one's "interests lay."[164] This, in turn, satisfies our "inner compulsion to be doing just what we were put here to do." In this sense, being able to live authentically is an intrinsic value in its own right. Even more strongly, philosopher Lawrence Becker proclaims that "autonomous human lives have a dignity that is immeasurable, incommensurable, infinite, beyond price."[165] Even Tolstoy recognized the power of self-expression in guiding human activities, writing, "In everything, in almost everything that I have written, I was guided by the need to bring together ideas linked among themselves, in order to achieve self-expression."[166]

Psychological research confirms what the philosophers have proposed about the importance of authenticity for meaning in life. Rebecca Schlegel and her colleagues at Texas A&M have demonstrated how various forms of authenticity and self-expression are connected to experiencing stronger meaning in life. In one study, Schlegel asked a group of students to provide as much detail as possible about their "true self," about "who you believe you really are."[167] Another group of students was asked to write about their "everyday self" as defined by how they actually behave in their daily life, and a third group of students was asked to write about the campus bookstore. After the writing task, the students were then asked to rate their meaning in life.

The researchers were interested in how much detail the participants provided in their various essays, their assumption being that the more detailed description one provides about one's true self, the more likely one is authentically in touch with that sense of self. Not surprisingly, for those people writing about their actual, everyday self or about the campus bookstore, the amount of detail didn't have any connection with their sense of meaning in life. However, when people wrote about their true selves, the more detailed the essay, the more the person on average experienced meaning in life. This and other research conducted by Schlegel and others demonstrates how being in touch with one's authentic self is connected to a higher sense of meaning in life.[168] Here, Schlegel's empirical research backs up what great humanistic psychologists like Carl Rogers and Abraham Maslow proposed decades ago and self-determination theory more recently professed: Autonomy is a fundamental human need, and thus we find intrinsic value in a chance to live our lives authentically. Being able to express oneself is part of a fully lived life, and such self-actualization can make our lives feel truly worth living.

This quality of our experience—how much we feel our choices and actions are autonomous and authentic instead of being controlled by external pressures—has important implications for human wellness, growth, and integrity, as research within self-determination theory has demonstrated in hundreds of scientific articles showing the importance of autonomy in various life domains ranging on everything from education and parenting to sports and physical exercise to weight loss and quitting tobacco smoking to the workplace and even dental hygiene.[169] The overall conclusion is that a sense of autonomy matters for both behavioral outcomes—autonomously motivated people are more likely to reach their goals—and wellness outcomes, such as being more satisfied with one's life, experiencing more positive feelings, and feeling more vital and energized.

To truly understand autonomy, it's crucial to realize that it's not the same as individualism. Autonomy is namely not only important to well-being in Western countries, where individualism has become a cultural norm, but also in countries like China, South Korea, Turkey, Russia, and Peru.[170] Individualism emphasizes human separateness and self-assertion, nonreliance on others, and prioritizes individual preferences and values over collective preferences and values. Autonomy, in contrast, is about feeling that one's actions and choices are self-chosen and not externally controlled. This means that one can autonomously endorse collective values. In other words, I can autonomously choose to help you; I can autonomously choose to prioritize the well-being of my own children over my own well-being. Accordingly, researcher Valery Chirkov showed that while students in Russia and South Korea viewed their cultures as being more collectivistic than students in the US, there were differences in all three countries about how autonomously people endorsed these cultural values.[171] In his research, some American students felt that individualism had gone too far in their country while others fully endorsed the same individualism. Similarly, some South Korean students had fully internalized the collective values of their country and were autonomously committed to them while others yearned for more individualistic values. And this—how autonomously they endorsed the cultural values of their own country—predicted the students' well-being in all three countries. So whether or not the country is collectivistic or individualistic, a sense of autonomy within that culture matters for one's personal well-being. It's when we're able to stay true to ourselves and live according to our values and interests that we truly come alive. As Deci and Ryan put it, "the fullest representations of humanity show people to be curious, vital, and self-motivated. At their best, they are agentic and inspired, striving to learn; extend themselves; master new skills; and apply their talents responsibly."[172] In essence, our humanity is best realized when we are able to pursue self-selected activities and live authentically in touch with who we really are and what we really value. Being able to do that makes life worth living.

DARE TO CARE: HOW TO BE AUTONOMOUSLY ALTRUISTIC

Have you ever swum through icy waters fully clothed and without a life jacket to rescue an unconscious woman you've never met before? Firefighter Jack Casey has. In the course of two years he responded as a volunteer to more than five hundred emergency calls, risking his own personal safety to save several people from life-or-death catastrophes. In addition to being a member of the local rescue squad, he spent three hours a week teaching a Red Cross course in first aid and took people backpacking through an outdoor program he initiated a few years before. Sociologist Robert Wuthnow, who interviewed Jack Casey for his book *Acts of Compassion,* saw in him a truly selfless American hero who wanted to be a bedrock of stability and safety for others.[173] At the same time Jack described himself as a person "who liked to be relatively independent of other people." He prided himself on being a rugged individualist who did what he wanted, when he wanted, and was free to think his own thoughts, and disagreed with others when it was called for. For Jack, and many others, freedom was a core American value.

To Wuthnow, Jack represents what he called an American paradox: he's more autonomous and independent than the average person yet he's also more caring of others, too. Is he an individualist or an altruist? The answer is he's both. In fact, it's his autonomy and independence from social norms that allowed Jack Casey to be sensitive to what he himself wants to do in life. And what he wants to do is to help others. Not because he *has to* but because he truly *wants to*.

Are we able to make independent choices or are we so weak and dependent on others that we let someone else run our lives? This is the question of individualism. Do we care only about ourselves or do we truly care about others as well? This is the question of altruism. But there's a third question: Can a person make independent choices while simultaneously recognizing that one's sympathy and care extends beyond oneself to other members of the social fabric? This is the essence of relationalism, which is a kind of individualistic altruism, and, in this sense, Jack Casey is clearly a relationalist: he makes his own choices and choses his own direction in life, which leads him to devote a significant amount of his time to helping the people around him.

Thus it's possible to be an independent altruist, someone who makes life decisions that lead to autonomously wanting to help those in need. As long as the choice is your own, you're still independent. Today, many people suppress their more altruistic instincts in order to live up to the selfish norms of our times. As social animals, it's in our nature to care about other people, but people ignore this altruistic instinct within themselves in order to fit into the culture that praises egoistic choices as the "rational" choices. In our culture, selfishness is all too easily equated with being smart while those who help others are seen as being foolish. Because of this cultural sentiment, people don't dare to be unselfish as they fear people may mock them for not being able to take care of their own interests. It takes courage to admit that one did something for someone else without any self-interest in mind. Perhaps, unsurprisingly then, it's become a modern-day paradox that independent-minded people may be more altruistically inclined than less independent people. Why? Because independent-minded people also tend to be more autonomous and, as such, are able to stand outside the cultural norms. And, in bucking the system, they're often better at expressing their more other-oriented, relational, and altruistic tendencies.

AUTONOMY IN EXTREMELY CONSTRAINED SITUATIONS

> "Between stimulus and response, there is a space. In that space is our power to choose our response. In our response lies our growth and our freedom."
>
> —STEPHEN COVEY,
> *First Things First*, 1995

Even in the most constrained surroundings, we still have a seed of freedom within our power: the freedom to choose how we interpret the situation and how we react. Sartre wrote that "once freedom lights its beacon in a man's heart, the gods are powerless against him."[174] Life has its constraints. A prisoner, for example, cannot will himself out of prison. But even the prisoner can choose how he reacts to his imprisonment. Having spent nine months as a prisoner of war during the Second World War himself, Sartre had some personal experiences to back up his conclusion. Viktor Frankl expressed a similar sentiment when he wrote: "We who lived in concentration camps can remember the men who walked through the huts comforting others, giving away their last piece of bread. They may have been few in number, but they offer sufficient proof that everything can be taken from a man but one thing: the last of the human freedoms—to choose one's attitude in any given set of circumstances, to choose one's own way."[175]

Life has its constraints and yet we still have the freedom to choose how we react to constraining situations. If our happiness and our sense of meaning is dependent on external circumstances, it is by necessity fragile because we can't control the external world. Accidents happen, people you love die, and some degree of tragedy is a part of every life. What we do have a degree of control over, however, is our own reaction to the external world. For Frankl, and for many others, it's what we do within the space of our own way of reacting—the emotional and psychological choices we make—that help set the course for our own personal liberation and give us the hope and endurance needed during moments of suffering, both extreme and mundane.

The ancient Stoics saw *apatheia*, literally translated as *without passion*, as a key life ideal. It's a state of mind where a person may observe whatever happens in life without judgment—to let it happen, to contemplate it, even to react to it in an appropriate way, but to not be overwhelmed by it. Apatheia is about being able to retain a certain distance between whatever happens in the world and how one reacts to it, to not allow it to get under the skin. When we're able to attain such a peace of mind, no external happenstance can disturb our inner calmness. Of course, attaining this state of mind is no easy feat. That's why Stoics included a whole program of various exercises by which a person could slowly mold one's mind toward apatheia. It is thus possible to experience a certain degree of freedom in even the most constrained situations, but getting to this degree of liberation from external circumstances requires a state of mind only attained through significant work.

Importantly, while taking steps toward this calmness of mind can help individuals in various intolerable situations, we shouldn't misinterpret it as an excuse to not make societies more supportive of autonomy. If research on self-determination theory has taught us anything, it's that the more autonomy-supportive schools, communities, and workplaces are, for example, the better people are able to realize themselves and the better their sense of well-being.[176] People can't thrive in environments where they are tightly controlled and forced to do things they wouldn't want to do. If a workplace culture is oppressive, the solution isn't to teach employees mindfulness techniques to help them better cope with oppression. The solution is to redesign the organization to allow more room for individual freedom and self-expression. As a society and as citizens, we should strive to build contexts, organizations, and governments that allow the individual to experience authenticity and autonomy in their actions, not only in their thoughts.

MASTER YOUR 12 POTENTIAL

"There is an irrepressible tendency in every man to develop himself according to the magnitude which Nature has made him of; to speak out, to act out, what nature has laid in him. This is proper, fit, inevitable; nay it is a duty, and even the summary of duties for a man. The meaning of life here on Earth might be defined as consisting in this: To unfold your *self*, to work what thing you have the faculty for."

—THOMAS CARLYLE,
Heroes, Hero-Worship, and the Heroic in History, 1840

There's a certain beauty in excellence. Whatever the field—be it modern dance, basketball, political oratory, gastronomy, or simultaneously juggling and solving three Rubik's cubes—we watch in awe if something is done with excellence. We can't help but admire it. Having played amateur-level soccer throughout my life, seeing the mastery with which players like Lionel Messi, Cristiano Ronaldo or Marta Vieira da Silva move the ball makes me break into spontaneous cheers. When reading, I sometimes stop to appreciate a particularly well-written sentence. If someone is the world's best in something—whatever the field—we immediately hold that person in high regard. Before the current global era, being best at a local level was enough to win the admiration of people around you. In his eighties, my grandfather could remember the names of the track and field athletes who had won the county championships of his youth.

Is there a virtue in excellence? The ancient Greeks thought so. For them, excellence *was* virtue. *Aretê*, the Greek word for virtue, was used in Homeric poems to describe excellence of any kind.[177] A fast runner displays the *aretê* of his feet, a big and strong swordsman displays the *aretê* of his physical strength. In these ancient heroic societies described in stories by Homer, or in Icelandic sagas, or Irish tales of the Ulster heroes, man's ultimate obligation was to fulfill the duties of whatever role society prescribed him. Whatever helped him fulfill his role was considered virtuous and whatever made him fail was a vice. In analyzing the morality of these heroic societies, philosopher Alasdair MacIntyre wrote, "the virtues just are those qualities which sustain a free man in his role and which manifest themselves in those actions which his role requires."[178] In other words, for a hero to fulfill his duties, sometimes he needs physical strength and sometimes he needs cunning. Virtue was found in possessing the required excellence.

There is a certain inherent meaningfulness in mastery and excellence in and of itself. As philosopher John Rawls stated, "Human beings enjoy the exercise of their realized capacities (their innate or trained abilities), and this enjoyment increases the more the capacity is realized, or the greater its complexity."[179] Being able to excel in something is valuable as such; we derive a sense of meaningfulness from having activities in our lives where we can experience competence, mastery, and skillfulness. How else can we explain the fact that humans put in enormous hours of work to master skills that have no point beyond the thing itself? There are, of course, the more established fields like professional sports or the arts into which billions of dollars are poured because we don't seem to get tired of watching people excel whether in competition, performance, or artistry. But when we think about the various trivialities in which people specialize—hula hooping while waterskiing, air guitar playing, chess boxing, or perfecting the art of coffee brewing[180]—here's the naked truth: people enjoy and derive meaning from being excellent in almost anything.

Competence, as the self-determination theory suggests, is a basic psychological need. As creatures of evolution, it's in our best interest to develop a wide range of skills because we never know which one might save our lives. If nothing poses a direct threat, the best use of our time might be to develop or hone some skills that could come in handy later. Idleness is a worse survival strategy than skill-acquiring. Accordingly, evolution has equipped us with a keen motivation to seek opportunities to learn new skills and derive satisfaction when we feel we're getting better at being able to do something well. Learning and personal growth are great sources of satisfaction and engagement and make us feel that our life is progressing. And when we're able to immerse ourselves in an activity that we master, we sometimes get so absorbed into it that we seem to forget the whole outside world. Psychologist Mihaly Csikszentmihalyi coined the term *flow* to describe this deep sense of absorption that athletes or artists sometimes get into when concentrating on a challenging-enough project.[181] In this state of absorption, one concentrates on the task at hand with complete attention and energy, moving effortlessly to attend to it, the conscious and unconscious mind in harmony. Csikszentmihalyi saw this as a kind of optimal state of inner experience and realized that the experience itself was so enjoyable that "people will do it even at great cost, for the sheer sake of doing it."[182] People seek challenges where they are able to utilize their skills, not just to accomplish something afterward but because, in moments of complete absorption, they often feel most alive. Indeed, a survey of four thousand Americans showed that some four hundred of them spontaneously mentioned "life struggles" as being a source of meaning.[183] Subsequently, I see that mastery and challenges can provide one with an important source of meaning in life.[184]

Again, it's important to realize that mastery as a source of meaning doesn't have to be derived from something extraordinary.

Climbing Mount Everest or sailing solo around the world surely might give oneself a strong sense of mastery—at the very least they're admirable endeavors and make for great conversation at cocktail parties. However, mastery can be felt in our everyday lives. I feel a sense of mastery when I play ball with my kids, never mind they're half my size, or when I'm able to climb the few steep hills on my daily bike commute. I feel a slight bang of accomplishment and mastery when, through my Tetris-honed skills, I'm able to fit the last dish into the dishwasher. Even cleaning the house can be a source of mastery and enjoyment for some (or so I've heard). These might sound like small and mundane moments, but it's both the big and small achievements and moments of mastery from which the sense of meaningfulness in our everyday lives is derived.

Of course, besides being able to savor these mundane moments of mastery, it's also great to actively seek out opportunities to develop skills in areas where we can reach an even stronger sense of mastery. At work, for example, it makes sense to be aware of the skills needed to advance to more challenging roles, and to set up systematic ways to train and hone that specific skill set. And outside of work, be sure to have a few enjoyable hobbies in which you feel you're able to experience a sense of mastery or learning.

SELF-ACTUALIZATION THROUGH MASTERY AND INTEREST

When we engage in an activity and feel a deep sense of mastery and interest in it, the activity itself becomes a source of self-actualization. What do I mean when I talk about self-actualization? Today, the term gets bandied about, and has been used to describe everything from yogic consciousness to New Age belief systems. I, however, see self-actualization as something more simple and mundane: self-actualization is about the simultaneous satisfaction of two needs: autonomy and competence. If you're able to find an activity that you'd like to do and also feel competent and capable that you can do it, you'll experience a sense of personal fulfillment, or self-actualization, in its pursuit. If one or the other is lacking, you don't feel a sense of self-actualization. For example, no matter how excited you get about a particular activity, if you don't feel that you are accomplishing anything by doing it or feel stuck, you get less and less enthusiastic doing it. When this happens, and no learning or growth seems to take place, it's next to impossible to sustain motivation and usually leads you to conclude that this particular activity may not be the thing for you. But the opposite is also true: no matter how skilled you are at an activity, if you're not invested in it, it doesn't feel personally fulfilling nor like a path to self-actualization. Talent can quickly become a trap if it's in an area you have little interest in. Fortunately, you can also identify activities that you find interesting and autonomously motivating, but in which you are also able to display a degree of mastery—especially if you're willing to put in the hours of practice. When interest and mastery meet in an activity, a sense of self-actualization occurs.

CONNECTING WITH OTHERS AND CONNECTING WITH YOURSELF: TWO PATHS TOWARD A MORE MEANINGFUL LIFE

There may be many paths to a more meaningful life. But I see that on a general level, the two key pathways to meaningful living include connecting to yourself and connecting to others. You connect to yourself through authenticity and mastery, and you connect to other people through close relationships and making positive contributions. Connecting to yourself leads to personal fulfillment. It's about authenticity, about being able to make autonomous choices to pursue your interests, and to express who you really are in both words and actions. Instead of conforming to external expectations, you're able to be true to yourself. It's also about mastery in that you're learning more about yourself and, in all matters of self-growth, can apply your newfound self-knowledge, skills, and strengths to everyday life.

> "One can live magnificently in this world, if one knows how to work and how to love, to work for the person one loves and to love one's work."
>
> —LEO TOLSTOY, *Letter to Valerya*, 1856

Connecting with others is about social fulfillment. It takes place through connecting with the people you care about, building good relationships, and being able to spend time with the people you love. But it's also about sensing that your life positively contributes to the lives of other people and that you're able to make a difference, however small. I've said it before and I'll say it again: through connecting and contributing, you make your life meaningful by being meaningful to others. This bit of advice, however, only covers the social fulfillment side of meaningful living. To further incorporate the idea of personal fulfillment, I've had to amend my one-liner. A philosopher walks into a bar and is asked to distill the meaning in life into one sentence. Nowadays my answer is this: meaning in life is about doing things meaningful to you (personal fulfillment) in a way that makes yourself meaningful to other people (social fulfillment).

How these two key pathways get built are, of course, up to you; they are wholly dependent on your personal interests, values, skills, and life situation. Only you can put in the work to connect the dots—autonomy, competence, relatedness, and benevolence—to the particularities and possibilities of your own life. Your compassionate, politically minded friend might use her oratory talent (self-actualization) to fight for a cause close to her heart (connecting with others). Your musically talented colleague could play guitar in a garage band (self-actualization), taking delight in the joy of jamming together with bandmates (connecting with others). A hospital janitor might enjoy the concrete results (self-actualization) that uphold the hygiene (connecting with others) of each patient's room. For many of us, parenting is a channel for both self-expression and contribution and the same applies for many people's hobbies and volunteer work. Of course, sometimes one aspect of your life might satisfy one need, while others may satisfy another. Having highly interesting yet lonely work can be compensated for by investing in relationships during leisure time. But no matter where you are in life, it's beneficial to think about how you can express yourself and contribute to the world at large.

A PERFECTLY MEANINGFUL MOMENT

A particularly meaningful moment in my life happened recently: my five-year-old son and I took our first biking trip together to a cafeteria by the sea a few miles from our home. I can recall every detail of our moment together—me drinking a cup of coffee; him quietly sipping from his juice box; the streaming sunlight; the faint smell of the sea—as if it happened earlier today. You could say that I simultaneously experienced the past, the present, and the future in the moments between my sips of coffee as I flashed backward in time, remembering my son's first steps as a baby while I also enjoyed the quiet relaxation following our bike ride and cast my mind forward to the future, envisioning all the other bike rides and coffees and quiet moments he and I had yet to share now that he was big enough to ride his own bike. My state of being was one of sheer delight and it sprung from the meaningfulness of the moment.

Truly at ease with myself, I gave in to the moment entirely and my son, perhaps graced with the kind of worry-free innocence that often accompanies childhood, did too. We mutually shared the space between us; it was a moment of a deep sense of belonging—a moment of love.

In that moment, all the key elements of meaningfulness were present. In being able to ride a bike I felt the pleasure of expressing myself through doing something I've always loved to do. At the same time, I shared in and enjoyed my son's newfound sense of mastery as much as he did. Through these experiences of self-expression and shared mastery, I felt connected to myself and to my son, resulting in a deep sense of belonging. I was also proud and happy as a father to have been able to offer this adventure to my son and took delight in the joy and excitement our mutual trip awakened in him, which enhanced my feeling of contribution. Past, present, and future. Self-expression, mastery, belonging, and contribution. What more could one ask for as regards meaning in life?

My story, of course, isn't exceptional. I'm sure you've experienced something similar in your own life with your loved ones— or with someone you've recently met. A moment like mine is a glimmer of existence when, for whatever reason, everything lines up—all the elements of meaningfulness are seemingly in sync with one another and you're fully immersed in your life. It's a moment where connecting with others and connecting with oneself merge together. It's personal and social fulfillment. It's meaning at its most essential.

Meaningfulness isn't something remote or rare. It's an experience that exists in many of our everyday moments in stronger or weaker form. Emily Esfahani Smith, author of *The Power of Meaning,* argues that meaning is "not some great revelation. It's pausing to say hi to a newspaper vendor and reaching out to someone at work who seems down. It's helping people get in better shape and being a good parent or mentor to a child."[185] These are tiny moments of social fulfillment, while small moments of personal fulfillment can take place when we dance like no one's watching or immerse ourselves in a book on our daily commute. They are meaningful here and now, but if we're able to connect them somehow to our past—being something you did together with your grandmother when you were kid—or to a valuable goal we have in the future, this can further enhance the meaningfulness of these moments. No matter where you are in life, it's important to stay connected to who you are and to those people, values, and interests that make your life meaningful, as Tolstoy learned during one of the lowest points of his life.

Suffering from a profound, debilitating existential crisis, Tolstoy dug deep and decided to clarify what truly mattered to him in his life. He emerged from his depressive funk with what he called the last "two drops of honey" that kept him anchored to this world: "love for family" and "love for writing."[186] In other words: connecting with others and connecting with oneself. What are your two drops?

Make Do with What You've Got

How grand your sense of meaningfulness needs to be is ultimately up to you. You set your standards for how much authenticity, mastery, belonging, and contribution is needed for you to experience your life as meaningful. For most of us, it's enough to experience an everyday level of contribution in order to experience a sense of positive impact. We all can't be Nelson Mandela or Martin Luther King Jr., and this is precisely what sets them apart from the rest of us. I once did observational research in a nursing home, and a scene from one of those days has stayed with me: Two elderly residents had taken over the job of folding the daily linen from the nursing staff. The women approached their task with concentration, engagement, and a sense of importance. Being residents, they had few opportunities to contribute, but here was a chance for them to give back and alleviate some work for the nurses who helped them so much. Folding linen isn't a difficult task in and of itself, and it's quite far from Mandela's accomplishments. Given their life circumstances, however, it was the perfect opportunity for them to satisfy their need for making a contribution. Through making themselves meaningful to the nurses they were able to enhance the meaningfulness of their own lives.

So set the standards of meaningfulness that are appropriate for your life and circumstances. If you're blessed with the opportunities to make great contributions, build deep relationships, or attain a world-class level of mastery in something you feel is your path to self-expression, then by all means set grand goals that stretch you to surpass yourself and attain the impossible. If you have a plethora of resources, be they financial, social, intellectual, or whatever, make the best out of them and give back at least as much as what you've received. If you are, however, in a tighter spot, make do with what you have and try to attain something feasible, whatever meaning is possible in your situation. Meaningfulness is as much about savoring as it is about attaining. Being able to recognize the small sources of meaningfulness already present in your life and the tiny ways you can further improve your relationships and your sense of contribution, mastery, and self-expression is already enough to make most of us feel that our lives are worth it.

LIVE YOUR LIFE NOT AS A PROJECT BUT AS A STORY

A final word of warning. Don't let the project of *making* your life meaningful get in the way of *experiencing* your life as meaningful. Modern Western culture has indoctrinated us into approaching our lives as projects. You're taught to set goals, make plans, aim high, and prioritize your efforts, all in the name of achieving the maximum outcome, that Holy Grail of Western life: success. When you approach life like it's a project, your life's value is dependent upon the project's success or failure. And given that these outcomes are often realized only in some remote future, the whole path to that point, which you may or may never reach, becomes tedious labor with no inherent value. At worst, it gets to a point where not suffering is taken as a sign that you're not working hard enough, as research scientist Emma Seppälä noted when observing the frantic achievement culture among Stanford University students.[187] The problem with projects is that they instrumentalize your life; it's no longer about living life but, instead, using life to attain something. It doesn't help much if you switch from striving for money, fame, and success to striving for happiness and meaning. While it may be wise to question the maximization of wealth, status, or career success, it means nothing if you've simply substituted one obsession for another. You're still using your life to attain something rather than embracing and living it. By keeping the focus on the end result, you fail to see the small, glimmering, everyday moments that actually make life meaningful.

Treat your life less as a project and more as a story, one that's wholly original to what you encounter, experience, witness, and express. Whatever happens to you, good or bad, self-selected or externally imposed—it's still part of the story. The chapters of your story include your various strengths and weaknesses, quirks and uniqueness.

A story isn't a competition, either. A story unfolds. It calls people to action; it begs its characters to make choices. As reflective creatures, we love a good story and for good reason: we use stories to teach lessons, moral and personal. We use them to entertain ourselves, but also to reflect on our world at large and to give it meaning. We look to stories to restore our sense of beauty and sanity in an often confusing and complex world. Certainly, there can be projects within a story; after all, big projects make great material for a good story. But projects are only part of the story. Don't let them dominate you, your worldview, or your sense of meaningfulness in life. In the end, the story of your life unfolds in the present moment, and the only thing you can strive for is, in John Dewey's words, the "enrichment of the present for its own sake."[188]

There's an old Eastern tale about a traveler walking peacefully on a steppe when he is suddenly overtaken by a tiger. Running for his life, the traveler comes to an edge of a cliff and jumps. To his horror he notices an enormous crocodile waiting for him at the bottom of the cliff, mouth open and ready to swallow him. In a desperate move, the traveler quickly clutches the twig of a wild bush growing along the cliff. He's caught between two horrible options: the tiger above and the crocodile below. Two mice begin to nibble the twig from which he hangs. He knows his death is inescapable.

Tolstoy used this tale to illustrate the condition of his life. In his existential crisis, he saw himself as that traveler, unable to enjoy anything life had to offer, because he could only focus on the mice and the crocodile.[189] But there's more to the Zen koan than what Tolstoy derived from it: instead of being obsessed with the inevitable death, the traveler focuses on what beauty there is still available in the present moment. Next to the twig are a few glistening strawberries, which he grabs with his other hand. Upon eating them, he thinks for himself, *So sweet they taste!*

Life may end one day. Every other day it doesn't. During those every other days we have opportunities to appreciate beauty, cultivate meaning, and find sweetness. A wonderful life is a life attuned to the small wonders of our everyday life. This idea is further expanded upon by Alan W. Watts, the famous Zen teacher, who compares life to music. He notes that in music one doesn't make the end of the composition the point of the composition. In playing a song, the one who plays it fastest doesn't win. What is meaningful in music is not getting to the end but what happens during the moments when the music is played.[190] In his words, "We thought of life by an analogy—as a journey or a pilgrimage—which has a serious purpose at the end. The thing was to get to that end, success, or whatever it is, or maybe Heaven after you are dead, but we missed the point along the whole way. It was a musical thing, and you were supposed to sing or dance while the music was being played." One day the music will stop. What happens afterward no one knows. But there's no point in waiting for the silence. If you're reading this, then the music is still playing for you. So go out and dance.

EP

"Every atom in your body came from
...You are all stardust....The stars died

OG

TRUE

a star that exploded.

so that you could be here today."

—LAWRENCE KRAUSS, *A Universe from Nothing*, 2009

Humankind's amazing capability to love, celebrate, mourn, sing, dance, and dream emerging from a pile of oxygen, carbon, hydrogen, and a bunch of other atoms—that's a wonder worth appreciating. The more we appreciate human life as a random, improbable event, the more we should be thankful that every one of us was given a unique life to live. Human existence has value and meaning even in a universe devoid of any absolute value; in fact, *you* make existence valuable by assigning value to it.

Rather than contemplating the meaning *of* life, focus on the meaning *in* life.

Meaning in life isn't about life in general; it's about your life. It's about how you can experience your unique existence as meaningful and worth living. And experiencing your unique existence as meaningful is simpler than you might think: connect with yourself by pursuing activities and goals meaningful to you; seek places to grow and utilize your areas of mastery; connect with others by cultivating intimate relationships; and do something good for others. These four sources of meaning might not sound too revolutionary, but that's exactly their strength. They are the cornerstones of a meaningful existence that you—and everyone else—already recognize as valuable. Stop chasing the phantoms of past ages. Stop longing for meaning imposed on life from above. Stop letting others set the standards for your life. Instead, concentrate on making your own life—and the lives of those you care about—more meaningful. This might sound like simple advice, but therein lies the path to meaningful living. As Camus wrote: "A single truth, if it is obvious, is enough to guide an existence."[191]

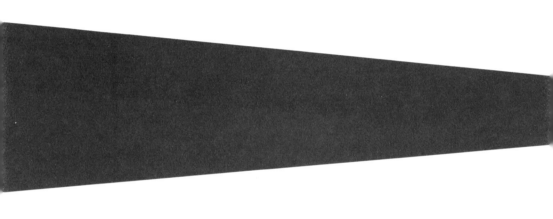

"Throw roses into the abyss
and say: 'Here are my thanks
to the monster for not knowing
how to swallow me alive.'"

—FRIEDRICH NIETZSCHE,
Nachgelassene Fragmente, circa 1883

I t takes a village to write a book. Without the various conversations I've had with a broad group of people—researchers, philosophers, friends, family, strangers on hotel verandas—or the thought-provoking articles and books I've had the pleasure of reading over the years, this book wouldn't exist. So, rather than it being solely my creation, this book is a distillation of the wisdom I've gathered from those conversations among friends, family, and strangers alike and from the writings of authors and fellow colleagues I consider it a privilege to have learned from. Mentioning everyone would be impossible, but here are the few people who have most directly helped in making this book happen.

First, I want to thank Signe Bergstrom, who provided invaluable assistance in transforming grand philosophical ideas and theses into a readable set of insights. She really put in the effort to think together with me how to arrange and present the ideas, and how to write about them in an engaging way. Writing the book with her on my team proved to be a great learning experience, and I warmly thank her for her input. Additionally, at Harper Design, I want to thank Elizabeth Viscott Sullivan and Marta Schooler for believing in this book and for their great support in putting it together. I'd also like to thank Roberto de Vicq de Cumptich for his beautiful design work. Furthermore, I want to especially thank my agents, Elina Ahlbäck and Rhea Lyons and their team, who have done a tremendous job in turning my book idea into a reality by perfecting my proposal, supporting me, and persistently spreading the word about it to get it published.

Various people either commented on parts of the manuscript or had meaningful conversations with me while I worked on it, the insights of which made their way into this book in one form or another. Thank you for providing new ideas, challenging my insights, and for giving me valuable feedback that helped improve my arguments: Ed Deci, Adam Grant, Antti Kauppinen, Laura King, Dmitry Leontiev, Jani Marjanen, Thaddeus Metz, Gregory Pappas, Holli-Anne Passmore, Anne Birgitta Pessi, Richard Ryan, Esa Saarinen, Emma Seppälä, Kennon Sheldon, Michael Steger, Jaakko Tahkokallio, Wenceslao Unanue, and Monica Worline as well as the participants of the Moral and Political Philosophy Research Seminar at University of Helsinki and participants of the Meaning of Life Conference at Harvard University. I especially want to highlight the contributions of Esa Saarinen and Richard Ryan, who have been key intellectual mentors to me in my work as a researcher. Thanks also to my friends and colleagues for the journey we've gone on together and for helping me grow into who I am today: Lauri, Karkki, Tapani, Timo, and other people from Filosofian Akatemia; my fellow students of philosophy Eetu, Hanna, Johanna, Kalle, Markus, Matti, Sanna, and Reima; my reading group and floorball friends Akseli, Antti H., Antti T., Janne, Jouni, Juha, Mikko, Olli, Timur, Touko, and Ville; and my siblings and their spouses, Eero, Tiia, Anna, and Tomi!

I also want to thank John Dewey, my philosophical mentor from a century ago, whose writings have provided a solid basis upon which to build my own insights, and on whose shoulders I am proud to stand.

Finally, I want to thank my parents, Heikki and Maarit, for providing an encouraging and supportive environment in which I grew. Having a big bookshelf and the time to read stories to them are the greatest gifts one can give to one's children. This is also the gift I want to provide my own children, Vikkeri, Roki, and Tormi. Together with my spouse, Piret, whom I want to thank for her support throughout the project and for giving me this gift of having a family, they are key to the meaningfulness in my own life!

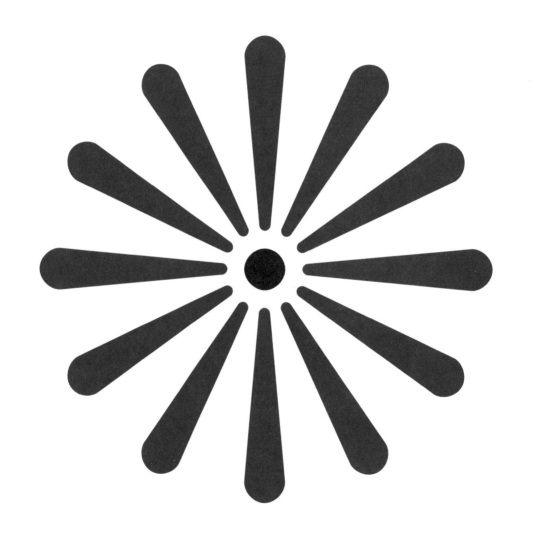

1 Roy F. Baumeister & Kathleen D. Vohs, "The Pursuit of Meaningfulness in Life," in *Handbook of Positive Psychology*, eds. Charles R. Snyder & Shane J. Lopez, 608–618 (New York: Oxford University Press, 2002), 613.

2 Lisa L. Harlow, Michael D. Newcomb & Peter M. Bentler, "Depression, Self-Derogation, Substance Use, and Suicide Ideation: Lack of Purpose in Life as a Mediational Factor," *Journal of Clinical Psychology* 42, no. 1 (1986): 5–21; Craig J. Bryan, William B. Elder, Mary McNaughton-Cassill, Augustin Osman, Ann Marie Hernandez & Sybil Allison, "Meaning in Life, Emotional Distress, Suicidal Ideation, and Life Functioning in an Active Duty Military Sample," *The Journal of Positive Psychology* 8, no. 5 (2013): 444–452. For a national-level comparison of rates of meaningfulness and rates of suicide, see Shigehiro Oishi & Ed Diener, "Residents of Poor Nations Have a Greater Sense of Meaning in Life Than Residents of Wealthy Nations," *Psychological Science* 25, no. 2 (2014): 422–430.

3 See Michael F. Steger, "Meaning and Well-Being," in *Handbook of Well-Being*, eds. Ed Diener, Shigehiro Oishi & Louis Tay (Salt Lake City, UT: DEF Publishers, 2018).

4 See Randy Cohen, Chirag Bavishi & Alan Rozanski, "Purpose in Life and Its Relationship to All-Cause Mortality and Cardiovascular Events: A Meta-Analysis," *Psychosomatic Medicine* 78, no. 2 (2016): 122–133 for a meta-analysis of ten prospective studies with a total of 136,265 participants.

5 Viktor Frankl, *Man's Search for Meaning* (New York: Washington Square Press, 1963), p. 164. The quote is originally from Nietzsche's book *Twilight of the Idols*.

[6] Alasdair MacIntyre, *After Virtue*, 3rd ed. (Notre Dame, IN: University of Notre Dame Press, 2007).

[7] See especially Christian Welzel, *Freedom Rising: Human Empowerment and the Quest for Emancipation* (New York: Cambridge University Press, 2013).

[8] Tim Kreiner, "The 'Busy' Trap," New York Times, June 30, 2012. https://opinionator.blogs .nytimes.com/2012/06/30/the -busy-trap/.

[9] Iddo Landau, *Finding Meaning in an Imperfect World* (New York: Oxford University Press, 2017), 205.

[10] This view of the human condition is derived from William James and John Dewey, two key figures in the pragmatist philosophical tradition. See Frank Martela, "Pragmatism as an Attitude," in *Nordic Studies in Pragmatism 3: Action, Belief and Inquiry— Pragmatist Perspectives on Science, Society and Religion*, ed. Ulf Zackariasson, 187–207 (Helsinki: Nordic Pragmatism Network, 2015), and the introduction of my second dissertation, Frank Martela, *A Pragmatist Inquiry into the Art of Living:Seeking Reasonable and Life-Enhancing Values within the Fallible Human Condition* (Helsinki: University of Helsinki, 2019).

[11] Albert Camus, *Myth of Sisyphus*, trans. Justin O'Brien (New York: Vintage Books, 1955).

[12] Todd May, *A Significant Life: Human Meaning in a Silent Universe* (Chicago: University of Chicago Press, 2015), ix.

[13] Robert N. Bellah, Richard P.

Madsen, William M. Sullivan, Ann Swidler & Steven M. Tipton, *Habits of the Heart* (Berkeley: University of California Press, 1985). The quotes below are from pages 20–22 and 76.

[14] Jean-Paul Sartre, *Existentialism Is a Humanism*, trans. Carol Macomber (New Haven: Yale University Press, 2007), 29.

[15] E.g., Angus Deaton, "Income, Health and Well-Being Around the World: Evidence from the Gallup World Poll," *Journal of Economic Perspectives* 22, no. 2 (2008): 53–72.

[16] Shigehiro Oishi & Ed Diener, "Residents of Poor Nations Have a Greater Sense of Meaning in Life Than Residents of Wealthy Nations," *Psychological Science* 25, no. 2 (2014): 422–430.

[17] These three dimensions of absurdity were identified by Joe Mintoff. See Joe Mintoff, "Transcending Absurdity," *Ratio* 21, no. 1 (2008): 64–84.

[18] Neil deGrasse Tyson, *Astrophysics for People in a Hurry* (New York: W. W. Norton & Company, 2017), 13.

[19] Thomas Nagel, "The Absurd," *The Journal of Philosophy*, 68, no. 20 (1971): 716–727, 717.

[20] Mintoff, "Transcending Absurdity."

[21] Leo Tolstoy, *Confession*, trans. David Patterson (New York: W. W. Norton & Co., 1983), 49.

[22] Carlin Flora, "The Pursuit of Happiness," *Psychology Today*, January 2009. Available online: https://www.psychologytoday .com/intl/articles/200901/ the-pursuit-happiness.

[23] Darrin M. McMahon, "From the

Happiness of Virtue to the Virtue of Happiness: 400 b.c.–a.d. 1780," *Daedalus* 133, no. 2 (2004): 5–17.

24 Geoffrey Chaucer, *The Canterbury Tales*, rendered into Modern English by J. U. Nilson (Mineola, NY: Dover Publications, 2004), 215.

25 See, e.g., Roy F. Baumeister, "How the Self Became a Problem: A Psychological Review of Historical Research," *Journal of Personality and Social Psychology* 52, no. 1 (1987), 163–176.

26 For a history of happiness, see especially Darrin M. McMahon, *The Pursuit of Happiness: A History from the Greeks to the Present* (London: Allen Lane, 2006).

27 In 1689, John Locke declared that the "pursuit of happiness" moved the human. John Locke, *An Essay Concerning Human Understanding* (London: Penguin Books, 1689/1997), 240. See also McMahon, "From the Happiness of Virtue to the Virtue of Happiness."

28 See, e.g., Charles Taylor, *The Ethics of Authenticity* (Cambridge, MA: Harvard University Press, 1991).

29 The exact definition of happiness is a big debate within Western philosophy and psychology, with some associating happiness with life satisfaction, others with the abundance of positive emotions and feelings, and still others building more complex accounts of happiness as an individual responding favorably, in emotional terms, to one's life. But for our present discussion, we don't need to settle on an exact definition of happiness, as what I am

going to say about it applies to all definitions of happiness that see it as some set of subjective feelings, emotions, or favorable responses to life. For discussion about the definition of happiness, see, e.g., Daniel M. Haybron, *The Pursuit of Unhappiness: The Elusive Psychology of Well-Being* (New York: Oxford University Press, 2008).

30 Luo Lu & Robin Gilmour, "Culture and Conceptions of Happiness: Individual Oriented and Social Oriented SWB," *Journal of Happiness Studies* 5, no. 3 (2004): 269–291.

31 Eric Weiner, *The Geography of Bliss* (New York: Hachette Book Group, 2008), 316–318.

32 Weiner, *Geography of Bliss*, 318.

33 Iris B. Mauss, Maya Tamir, Craig L. Anderson & Nicole S. Savino, "Can Seeking Happiness Make People Unhappy? Paradoxical Effects of Valuing Happiness," *Emotion* 11, no. 4 (2011): 807–815. See also Maya Tamir & Brett Q. Ford, "Should People Pursue Feelings That Feel Good or Feelings That Do Good? Emotional Preferences and Well-Being," *Emotion* 12, no. 5 (2012): 1061–1070.

34 Iris B. Mauss, Nicole S. Savino, Craig L. Anderson, Max Weisbuch, Maya Tamir & Mark L. Laudenslager, "The Pursuit of Happiness Can Be Lonely," *Emotion* 12, no. 5 (2012): 908–912.

35 For people having some degree of depressive symptoms, constant reporting of their happiness levels can be detrimental for their well-being. See Tamlin S. Conner & Katie A. Reid, "Effects

of Intensive Mobile Happiness Reporting in Daily Life," *Social Psychological and Personality Science* 3, no. 3 (2012): 315–323.

36 Many philosophers tend to see happiness and meaningfulness as two separate and fundamental values in human life. See Thaddeus Metz, *Meaning in Life: An Analytic Study* (Oxford: Oxford University Press, 2013), chapter 4; and Susan Wolf, "Meaningfulness: A Third Dimension of the Good Life," *Foundations of Science* 21, no. 2 (2016): 253–269.

37 In philosopher Robert Nozick's thought experiment about an experience machine providing all possible pleasures is the classic example of this. People are arguably not too eager to plug into such a machine. See Robert Nozick, *Anarchy, State, and Utopia* (Padstow: Blackwell, 1974), 42.

38 John F. Helliwell, Richard R. Layard & Jeffrey D. Sachs, eds. *World Happiness Report 2019* (New York: Sustainable Development Solutions Network, 2019). See also previous World Happiness Reports.

39 See Jakub Marian, "Number of Metal Bands per Capita in Europe," https://jakubmarian.com/number-of-metal-bands-per-capita-in-europe/.

40 John F. Helliwell et al., *World Happiness Report 2019*.

41 E.g., Max Haller & Markus Hadler, "How Social Relations and Structures Can Produce Happiness and Unhappiness: An International Comparative Analysis," *Social Indicators Research* 75,

no. 2 (2006): 169–216; Ronald Inglehart, Robert Foa, Christopher Peterson & Christian Welzel, "Development, Freedom, and Rising Happiness: A Global Perspective (1981–2007)," *Perspectives on Psychological Science* 3, no. 4 (2008): 264–285.

42 Jon Clifton, "People Worldwide Are Reporting a Lot of Positive Emotions," May 21, 2014. http://news.gallup.com/poll/169322/people-worldwide-reporting-lot-positive-emotions.aspx.

43 World Health Organization, *Global Health Estimates 2015: DALYs by Cause, Age, Sex, by Country and by Region, 2000–2015* (Geneva: World Health Organization, 2016).

44 Compare, for example, these two: Dheeraj Rai, Pedro Zitko, Kelvyn Jones, John Lynch & Richard Araya, "Country- and Individual-Level Socioeconomic Determinants of Depression: Multilevel Cross-National Comparison," *The British Journal of Psychiatry* 202, no. 3 (2013): 195–203; and Alize J. Ferrari, Fiona J. Charlson, Rosana E. Norman, Scott B. Patten, Greg Freedman, Christopher J. L. Murray, et al., "Burden of Depressive Disorders by Country, Sex, Age, and Year: Findings from the Global Burden of Disease Study 2010," *PLOS Medicine* 10, no. 11 (2013): e1001547.

45 See Brett Q. Ford, Phoebe Lam, Oliver P. John & Iris B. Mauss, "The Psychological Health Benefits of Accepting Negative Emotions and Thoughts: Laboratory, Diary, and Longitudinal Evidence,"

Journal of Personality and Social Psychology 115, no. 6 (2018): 1075–1092.

46 The relationship is logarithmic rather than linear, but whether there is a satiation point for life satisfaction is still debated among researchers. Positive affect has generally weaker relations with wealth and there the satiation point is more readily observed. See Daniel Kahneman & Angus Deaton, "High Income Improves Evaluation of Life But Not Emotional Well-Being," *Proceedings of the National Academy of Sciences*, 107, no. 38 (2010): 16489–16493; Eugenio Proto & Aldo Rustichini, "A Reassessment of the Relationship Between GDP and Life Satisfaction," *PLOS ONE* 8, no. 11 (2013): e79358; and Daniel W. Sacks, Betsy Ayer Stevenson & Justin Wolfers, "The New Stylized Facts About Income and Subjective Well-Being," *Emotion* 12, no. 6 (2012): 1181–1187.

47 Andrew T. Jebb, Louise Tay, Ed Diener & Shigehiro Oishi, "Happiness, Income Satiation and Turning Points Around the World," *Nature Human Behaviour* 2, no. 1 (2018): 33–38.

48 Jonathan Haidt, *Happiness Hypothesis: Finding Modern Truth in Ancient Wisdom* (New York: Basic Books, 2006), 89.

49 Chuck Palahniuk, *Fight Club* (London: Vintage Books, 2010), 149.

50 Statista, "Media Advertising Spending in the United States from 2015 to 2022 (in billion U.S. dollars)," March 28, 2019, https://www.statista.com/statistics/272314

/advertising-spending-in-the-us/.

51 Barry Schwartz, *The Paradox of Choice: Why More Is Less* (New York: HarperCollins, 2004).

52 Herbert A. Simon, "Rational Choice and the Structure of the Environment," *Psychological Review* 63, no. 2 (1956): 129–138. Simon contrasts the strategies of "satisfice" and "optimize."

53 See Nathan N. Cheek & Barry Schwartz, "On the Meaning and Measurement of Maximization," *Judgment and Decision Making* 11, no. 2 (2016): 126–146.

54 Samantha J. Heintzelman & Laura A. King, "Life Is Pretty Meaningful," *American Psychologist* 69, no. 6 (2014): 561–574.

55 *The Health and Retirement Study*, an ongoing longitudinal study of Americans over age 50 sponsored by the National Institute on Aging at the University of Michigan, http://hrsonline.isr.umich.edu/. These results are reported in Heintzelman & King, "Life Is Pretty Meaningful."

56 Rosemarie Kobau, Joseph Sniezek, Matthew M. Zack, Richard E. Lucas & Adam Burns, "Well-Being Assessment: An Evaluation of Well-Being Scales for Public Health and Population Estimates of Well-Being Among US Adults," *Applied Psychology: Health and Well-Being* 2 (2010): 272–297.

57 Oishi & Diener, "Residents of Poor Nations."

58 Fei-Hsiu Hsiao, Guey-Mei Jow, Wen-Hung Kuo, King-Jen Chang, Yu-Fen Liu, Rainbow T. Ho, et al., "The Effects of Psychotherapy on Psychological

Well-Being and Diurnal Cortisol Patterns in Breast Cancer Survivors," *Psychotherapy and Psychosomatics* 81 (2012): 173–182.

59 Heintzelman & King, "Life Is Pretty Meaningful," 567.

60 Sometimes people might be disillusioned and unreliable reporters of their own emotional states so we should take their reports with a grain of salt. But there is no evidence of bias so systematic that it would render subjective reports totally worthless, most of the time they probably are relatively accurate. See, e.g., OECD, *OECD Guidelines on Measuring Subjective Well-Being* (Paris: OECD Publishing, 2013). Be that as it may, I side with the psychologists in focusing on the experienced meaning of people rather than believing there to be some outside standard that can be used to judge the meaningfulness of people's existence no matter their own experience.

61 Jon H. Kaas, "The Evolution of Brains from Early Mammals to Humans," *Wiley Interdisciplinary Reviews: Cognitive Science* 4, no. 1 (2013): 33–45. Also see Joseph R. Burger, Menshian A. George, Claire Leadbetter & Farhin Shaikh, "The Allometry of Brain Size in Mammals," *Journal of Mammalogy* 100, no. 2 (2019): 276–283.

62 Yuval Harari, *Sapiens: A Brief History of Humankind* (New York: Harper, 2015).

63 See William A. Roberts, "Are Animals Stuck in Time?," *Psychological Bulletin* 128, no. 3 (2002): 473–489. Although see also William A. Roberts, "Mental Time Travel: Animals Anticipate the Future," *Current Biology* 17, no. 11 (2007): R418–R420, where evidence is reviewed for the capability of some animals to anticipate the future and have episodic-like memories. As with most things separating humans from animals, instead of a clear dichotomy, humans just have much more of what some animals have little of.

64 Antti Kauppinen, "Meaningfulness and Time," *Philosophy and Phenomenological Research* 84, no. 2 (2012): 345–377, 368.

65 Adam Waytz, Hal E. Hershfield & Diana I. Tamir, "Mental Simulation and Meaning in Life," *Journal of Personality and Social Psychology* 108, no. 2 (2015): 336–355, study 1.

66 Human sociality and living in large tribes most probably had much to do with why we developed the unique capacity for reflection as, e.g., Dunbar has argued. Also the need for justification most probably had much to do with the need to justify our actions to others as, e.g., Haidt has argued. See Robin I. M. Dunbar, "The Social Brain Hypothesis," *Evolutionary Anthropology: Issues, News, and Reviews* 6, no. 5 (1998): 178–190. And Jonathan Haidt, "The Emotional Dog and Its Rational Tail: A Social Intuitionist Approach to Moral Judgment," *Psychological Review* 108 (2001): 814–834.

67 See Frank Martela & Michael F. Steger, "The Meaning of Meaning in Life: Coherence, Purpose, and Significance as the Three

Facets of Meaning," *Journal of Positive Psychology*, 11, no. 5 (2016): 531-545.

[68] Erich Fromm, *Escape from Freedom* (New York: Avon Books, 1965), p. viii. The first edition of the book was published in 1941.

[69] Fromm, *Escape from Freedom*, xii.

[70] Michael F. Steger, Yoshito Kawabata, Satoshi Shimai & Keiko Otake, "The Meaningful Life in Japan and the United States: Levels and Correlates of Meaning in Life," *Journal of Research in Personality* 42, no. 3 (2008), 660–678.

[71] Charles Taylor, *A Secular Age* (Cambridge, MA: The Belknap Press of Harvard University Press, 2007). The description of the worldview of people living in the 1500s is mainly based on chapter 1 of Taylor's book.

[72] Taylor, *A Secular Age*, 42, based on Stephen Wilson, *The Magical Universe* (London: Hambledon & London, 2004).

[73] Max Weber, *The Sociology of Religion* (Boston: Beacon Press, 1971).

[74] There is, of course, a big leap in worldviews between belief in local spirits and one omnipotent creator God, but we have no space to go through the so-called axial revolution and how that transformed our worldviews.

[75] Aristotle, *Nicomachean Ethics*, trans. Robert C. Bartlett & Susan D. Collins (Chicago: University of Chicago Press, Chicago, 2012), 1094a:18–20.

[76] Aristotle, *Nicomachean Ethics*, 1106a:17–24.

[77] Note that Hochschild argues that this question was asked up to the twentieth century, while here I am arguing that meaning of life started to replace it in the nineteenth century. Joshua P. Hochschild, "What 'the Meaning of Life' Replaced," https://thevirtueblog.com/2017/12/18/what-the-meaning-of-life-replaced/.

[78] See the introduction by Kerry McSweeney and Peter Sabor in Carlyle, *Sartor Resartus*.

[79] The philosopher Wendell O'Brien argues that Carlyle's book is the earliest known piece of literature where the phrase "meaning of life" occurs. *Oxford English Dictionary* similarly gives Carlyle's book as the earliest example of the phrase. My own searches and consultations with a few specialists have not been able to find any previous occasion when somebody would have used the phrase "meaning of life." I would love to be proven wrong, but at this point I propose that the phrase was probably coined in English by Thomas Carlyle in 1833–1834, inspired by German romantics using the phrase "der Sinn des Lebens" a few decades earlier. See Wendell O'Brien, "The Meaning of Life: Early Continental and Analytic Perspectives," in *Internet Encyclopedia of Philosophy* (2014). Retrieved from http://www.iep.utm.edu/mean-ear/.

[80] Thomas Carlyle, *Sartor Resartus* (Oxford: Oxford University Press, 1987), 3.

[81] Carlyle, *Sartor Resartus*. The quotes in this paragraph are from pages 87, 89, 127, 140, 149.

82 Carlyle, *Sartor Resartus*, 211.

83 Søren Kierkegaard, *Either/Or*, trans. Howard V. Hong & Edna H. Hong (Princeton, NJ: Princeton University Press, 1987), 31.

84 Arthur Schopenhauer, *On Human Nature*, trans. Thomas Bailey Saunders (New York: Cosimo, 2010), 62. The phrase "Sinn des Lebens" appears also at least once in Schopenhauer's magnum opus *Die Welt als Wille und Vorstellung*.

85 Leo Tolstoy, *Tolstoy's Diaries, Volume I, 1847–1894*, ed. R. F. Christian (London: The Athlone Press, 1985), 191.

86 Tolstoy, *Confession*, 33–34. Note that this particular phrasing is from the translation by Louise and Aylmer Maude used by Antony Flew, "Tolstoi and the Meaning of Life," *Ethics* 73, no. 2 (1963), 110–118.

87 Tolstoy, *My Confession*, in *The Meaning of Life*, 2nd ed., ed. E. D. Klemke, trans. Leo Wierner, 11–20. (New York: Oxford University Press, 200), 15.

88 Jaakko Tahkokallio, *Pimeä aika* (Helsinki: Gaudeamus, 2019).

89 In Kepler's "Letter to Mästlin" in 1595. Quoted in James R. Voelkel, *The Composition of Kepler's Astronomia Nova* (Princeton, NJ: Princeton University Press, 2001), 33.

90 This, and other facts about the history of atheism in this paragraph, come from Gavin Hyman, *A Short History of Atheism* (New York: I. B. Tauris & Co., 2010), 3–7. He notes that among the ancient Greek and Roman thinkers we can sometimes find what he calls "soft" atheism. It was a form of freethinking, where some exceptional individuals proposed intellectual theories that saw the God's role in new ways or even denied much role for them. But these thinkers were few, typically did not reject religious practice, and didn't involve a resolute denial of a transcendent realm altogether.

91 Hyman, *A Short History of Atheism*, 7.

92 Tolstoy, *My Confession*, in *The Meaning of Life*, 19.

93 Alain de Botton, "How Romanticism Ruined Love," July 19, 2016, https://www.abc.net.au /religion/how-romanticism -ruined-love/10096750.

94 Douglas Adams, *The Hitchhiker's Guide to the Galaxy* (New York: Ballantine Books, 2009), 161.

95 Hyman, *A Short History of Atheism*, analyzes on pages 20–26 the transition taking place in the Descartes book.

96 See, e.g., MacIntyre, *After Virtue*.

97 Baumeister, "How the Self Became a Problem."

98 Hyman, *A Short History of Atheism*, xvi–xvii, notes that the world as something to be mastered and controlled and a sense of progress are key defining features of the modern worldview.

99 Furthermore, the peculiar European political situation was favorable for new ways of thinking. In the centrally governed China, the emperor had the power to set the limits of what one could think aloud. The European elite was culturally united and thus constantly exchanging thoughts, but politically Europe

was divided into small city-states and kingdoms. This meant that for a freethinker pushing the boundaries of what one could think, there was always some prince or ruler somewhere who was more liberal and tolerant and to whose principality one could flee if things started to get too heated in one's present location. I feel that we should not underestimate this historically peculiar condition—cultural unity within decentralized politics—in enabling the rapid development of several innovations in thinking in fields ranging from philosophy to astronomy to politics.

100 Of the unaffiliated, roughly one-third self-identify as atheists or agnostics (31%), roughly one-third say that religion is unimportant to their lives (39%), while one-third (30%) sees religion at least somewhat important in their lives but don't identify with any particular religion. While people in this latter group might still believe in some kind of God or spirituality, their way of believing is thus not connected to any particular religious group, but rather they believe or don't believe in their own way. See Pew Research Center, *America's Changing Religious Landscape*, Pew Research Report, 2015.

101 Pew Research Center, *America's Changing Religious Landscape*.

102 The statistic about not believing in God comes from the International Social Survey Programme (ISSP) 2008, quoted in Ariela Keysar, & Juhem Navarro-Rivera, "A World of Atheism," in

The Oxford Handbook of Atheism, ed. S. Bullivant & M. Ruse, 553–585, (New York: Oxford University Press, 2013). The amount of religiously unaffiliated is from Pew Research Center, *The Future of World Religions: Population Growth Projections, 2010–2050*, Pew Research Report, 2015. As regards the religiously unaffiliated, China, Hong Kong, and North Korea have—along with Czech Republic and Estonia—more than half of their populations reporting as religiously unaffiliated in the Pew Research Center 2015 data.

103 Based on the percentage of people who outrightly believe there is no God compared to people who believe in God without doubts in ISSP 2008 survey. See Figure 36.8 in Keysar & Navarro-Rivera, "A World of Atheism," 577.

104 Robert D. Putnam & David E. Campbell, *American Grace: How Religion Divides and Unites Us* (New York: Simon & Schuster), 4.

105 Taylor, *A Secular Age*.

106 Putnam & Campbell, *American Grace*, 6.

107 Carlyle, *Sartor Resartus*, 147.

108 This is how psychologists define meaning in life. Note that for philosophers, even the question of "meaning in life" typically refers to the objective meaningfulness of a particular life—whether that life is "really" meaningful—rather than to the person's subjective sense of meaningfulness.

109 Note that many analytic philosophers are nowadays still objective naturalists, believing that some

form of objectivism can be saved even if we accept naturalism. See, e.g. Metz, *Meaning in Life*, and Antti Kauppinen, "Meaningfulness," in *The Routledge Handbook of Philosophy of Well-Being*, ed. G. Fletcher, 281–291 (Abingdon, UK: Routledge, 2016). Unfortunately, here is not the time or space to discuss the merits and shortcomings of these attempts.

110 This sentiment is very much at the heart of Deweyan pragmatist philosophy. See especially Gregory Pappas, *John Dewey's Ethics: Democracy as Experience* (Bloomington: Indiana University Press, 2008).

111 See, e.g., the essays "Pyrrhus and Cineas" and "Introduction to an Ethics of Ambiguity" in Simone de Beauvoir, *Philosophical Writings*, ed. M. A. Simons (Urbana: University of Illinois Press, 2004).

112 Beauvoir in the essay "Introduction to an Ethics of Ambiguity," trans. by Marybeth Timmermann. In Beauvoir, *Philosophical Writings*, 291.

113 Beauvoir, *Philosophical Writings*, 293.

114 I talk about this idea of moral growth and its roots in pragmatism especially in Frank Martela, "Is Moral Growth Possible for Managers?," in *Handbook of Philosophy of Management*, ed. Cristina Neesham & Steven Segal. Advance online publication, doi:10.1007/978-3-319-48352 -8_18-1.

115 Frankl, *Man's Search for Meaning*, p. 112. Note that this quote is from the 2006 Beacon Press edition of the book.

116 Pappas, *John Dewey's Ethics*, 152.

117 John Dewey, *Human Nature and Conduct* (New York: Henry Holt and Company, 1922), 196.

118 Aristotle, *Nicomachean Ethics*, 1096b:3–4.

119 Samuel Beckett, *Waiting for Godot* (New York: Grove Press, 1954), 80.

120 Tim Urban, "The Tail End," *Wait But Why*, December 11, 2015, https://waitbutwhy.com/2015/12 /the-tail-end.html.

121 *Ferris Bueller's Day Off*, Broderick, Matthew. Directed by John Hughes. Los Angeles: Paramount Pictures, 1986.

122 This distinction between idiosyncratic sources of meaning and universal sources of meaning resembles Calhoun's distinction between reasons-for-anyone and reasons-for-me. See Cheshire Calhoun, *Doing Valuable Time: The Present, the Future, and Meaningful Living* (New York: Oxford University Press, 2018).

123 John Dewey, *Theory of Valuation* (Chicago: University of Chicago Press, 1939).

124 I develop this theme more in my article Frank Martela, "Moral Philosophers as Ethical Engineers: Limits of Moral Philosophy and a Pragmatist Alternative," *Metaphilosophy* 48, no. 1–2 (2017): 58–78.

125 Edward L. Deci & Richard M. Ryan, "The 'What' and 'Why' of Goal Pursuits: Human Needs and the Self-Determination of Behavior," *Psychological Inquiry* 11, no. 4 (2000), 227–268; Richard M. Ryan & Edward L. Deci, *Self-Determination Theory: Basic Psychological Needs in Motivation, Development, and Wellness*

(New York: Guilford Press, 2017). See also my brief introduction to self-determination theory in Frank Martela, "Self-Determination Theory," in *The Wiley-Blackwell Encyclopedia of Personality and Individual Differences: Vol. I. Models and Theories*, ed. Bernardo J. Carducci & C. S. Nave (Hoboken, NJ: John Wiley & Sons, in press).

126 Deci & Ryan, "The 'What' and 'Why' of Goal Pursuits," 229.

127 Deci & Ryan "The 'What' and 'Why' of Goal Pursuits"; Ryan & Deci, *Self-Determination Theory*.

128 Frank Martela & Richard M. Ryan, "The Benefits of Benevolence: Basic Psychological Needs, Beneficence, and the Enhancement of Well-Being," *Journal of Personality* 84, no. 6 (2016), 750–764.

129 In particular, while the frustration of the benevolence doesn't seem to lead to ill-being in the same sense as the frustration of the three established needs does, the satisfaction of the need for benevolence seems to lead to well-being and meaningfulness similarly to the three established needs. Instead of a basic psychological need, perhaps it is a kind of enhancement need. See Martela & Ryan 2019: "Distinguishing Between Basic Psychological Needs And Basic Wellness Enhancers: The Case of Beneficence as a Candidate Psychological Need." *Motivation and Emotion*, advance online publication. 10.1007/511031-019-09800.

130 See especially Frank Martela, Richard M. Ryan & Michael F. Steger, "Meaningfulness as Satisfaction of Autonomy, Competence, Relatedness, and Beneficence: Comparing the Four Satisfactions and Positive Affect as Predictors of Meaning in Life," *Journal of Happiness Studies* 19, no. 5 (2018), 1261–1282. The article also provides key references to the most important studies that I've seen that have examined these four as sources of meaning. See also Frank Martela & Tapani J. J. Riekki, "Autonomy, Competence, Relatedness, and Beneficence: A Multicultural Comparison of the Four Pathways to Meaningful Work," *Frontiers in Psychology* 9 (2018), 1–14.

131 A more philosophical treatment of this suggestion of building core values upon basic needs can be found in my article F. Martela, "Four reasonable, Self-Justifying Values—How to Identify Empirically Universal Values Compatible with Pragmatist Subjectivism," *Acta Philosophica Fennica*, 94 (2018), 101–128.

132 Israeli Professor Shalom Schwartz has probably done the most comprehensive work on human values around the world, with numerous studies from the 1990s to today involving people from more than a hundred different countries. Autonomy, competence, relatedness, and benevolence find somewhat corresponding values in his list of universal values shared by all cultures: self-direction, achievement, and caring-benevolence, and he readily acknowledges that the "values associated with autonomy, relatedness, and competence show a

universal pattern of high importance and high consensus." Ronald Fischer & Shalom Schwartz,
"Whence Differences in Value
Priorities? Individual, Cultural,
or Artifactual Sources," *Journal
of Cross-Cultural Psychology*, 42,
no. 7 (2011), 1127–1144. There is
a degree of dissimilarity between Schwartz's definition of
the relevant values and the basic
needs, which has to be taken
into account when comparing
his values and SDT's needs. But
I still feel that there is enough
conceptual similarity to make the
conclusion that Schwartz himself
makes: That the needs most
probably have almost universal
appeal. Schwartz notes that
benevolence could be divided
into two subtypes: dependability
which refers to relations with
friends and being responsible in
them, and caring, which is more
about being helpful. See Shalom
Schwartz, Jan Cieciuch, Michele
Vecchione, Eldad Davidov, Ronald Fischer, Constanze Beierlein
et al., "Refining the Theory of
Basic Individual Values," *Journal
of Personality and Social Psychology*, 103, no. 4 (2012), 663–688.

[133] Christopher P. Niemiec, Richard
M. Ryan & Edward L. Deci, "The
Path Taken: Consequences of
Attaining Intrinsic and Extrinsic
Aspirations in Post-College Life,"
Journal of Research in Personality
43, no. 3 (2009), 291–306.

[134] Kauppinen, "Meaningfulness and
Time," 364.

[135] Roy Baumeister & Mark Leary,
"The Need to Belong: Desire for
Interpersonal Attachments as a
Fundamental Human Motiva-
tion," *Psychological Bulletin*, 117,
no. 3 (1995), 497–529.

[136] Arthur Aron, Elaine N. Aron,
Michael Tudor & Greg Nelson,
"Close Relationships as Including
Other in the Self," *Journal
of Personality and Social
Psychology* 60, no. 2 (1991),
241–253.

[137] Yawei Cheng, Chenyi Chen, Ching
Po Lin, Kun Hsien Chou, & Jean
Decety, "Love Hurts: An fMRI
Study," *Neuroimage* 51, no. 2
(2010), 923–929.

[138] Maurice Merleau-Ponty, *Phenom-
enology of Perception*, trans. C.
Smith (London: Routledge, 2002
[1945]), 413.

[139] Nathaniel Lambert, Tyler F. Stillman, Roy F. Baumeister, Frank
D. Fincham, Joshua A. Hicks &
Steven M. Graham, "Family as
a Salient Source of Meaning in
Young Adulthood," *The Journal
of Positive Psychology* 5, no. 5
(2010), 367–376.

[140] Pew Research Center, "Where
Americans Find Meaning in
Life," November 20, 2018,
https://www.pewforum.org
/2018/11/20/where-americans
-find-meaning-in-life/.

[141] Lambert et al., "Family as a Salient
Source of Meaning"; Nathaniel
M. Lambert, Tyler F. Stillman,
Joshua A. Hicks, Shanmukh
Kamble, Roy F. Baumeister &
Frank D. Fincham, "To Belong
Is to Matter: Sense of Belong-
ing Enhances Meaning in Life,"
*Personality and Social Psychology
Bulletin* 39, no. 11 (2013): 1418–
1427; Martela et al., "Meaningful-
ness as Satisfaction of Autonomy,

Competence, Relatedness, and Beneficence."

142 Tyler F. Stillman, Roy F. Baumeister, Nathaniel M. Lambert, A. Will Crescioni, C. Nathan DeWall & Frank D. Fincham, "Alone and Without Purpose: Life Loses Meaning Following Social Exclusion," *Journal of Experimental Social Psychology* 45, no. 4 (2009), 686–694.

143 Robert D. Putnam, *Bowling Alone: The Collapse and Revival of American Community* (New York: Simon & Schuster, 2001). For a critique, see, e.g., Claude S. Fischer, "Bowling Alone: What's the Score?" *Social Networks* 27, no. 2 (2001), 155–167.

144 Jüri Allik & Anu Realo, "Individualism-Collectivism and Social Capital," *Journal of Cross-Cultural Psychology* 35, no. 1 (2004): 29–49, 34–35.

145 I heard the story for first time from Hintsa's friend and colleague Juha Äkräs. It is also told in Aki Hintsa & Oskari Saari's book, *The Core: Better Life, Better Performance*, trans. D. Robinson (Helsinki: WSOY, 2015), 196–198.

146 Leena Valkonen, "What Is a Good Father or Good Mother Like? Fifth and Sixth Graders' Conceptions of Parenthood / Millainen on Hyvä Äiti Tai Isä? Viides- Ja Kuudesluokkalaisten Lasten Vanhemmuuskäsitykset," (Jyväskylä: University of Jyväskylä, 2006), 42. Also quoted in Hintsa & Saari, *The Core*.

147 Frank Martela, "Meaningfulness as Contribution," *The Southern Journal of Philosophy* 55, no. 2 (2017): 232–256.

148 Frank Martela & Richard M. Ryan, "Prosocial Behavior Increases Well-Being and Vitality Even Without Contact with the Beneficiary: Causal and Behavioral Evidence," *Motivation and Emotion* 40, no. 3 (2016), 351–357.

149 See, e.g., Blake A. Allan, Ryan D. Duffy & Brian Collisson, "Helping Others Increases Meaningful Work: Evidence from Three Experiments," *Journal of Counseling Psychology* 65, no. 2 (2017), 155–165. Daryl R. Van Tongeren, Jeffrey D. Green, Don E. Davis, Joshua N. Hook & Timothy L. Hulsey, "Prosociality Enhances Meaning in Life," *The Journal of Positive Psychology* 11, no. 3 (2016), 225–236.

150 This section is based on a blog post I wrote that was originally published by *Scientific American Observations* entitled "Exercise, Eat Well, Help Others: Altruism's Surprisingly Strong Help Impact" (September 7, 2018), https://blogs.scientificamerican .com/observations/exercise -eat-well-help-others -altruisms-surprisingly-strong -health-impact/.

151 Ashley V. Whillans, Elizabeth W. Dunn, Gillian M. Sandstrom, Sally S. Dickerson & Kenneth M. Madden, "Is Spending Money on Others Good for Your Heart?" *Health Psychology* 35, no. 6 (2016), 574–583.

152 Stephanie L. Brown, Randolph M. Nesse, Amiram D. Vinokur & Dylan M. Smith, "Providing Social Support May Be More

Beneficial Than Receiving It: Results from a Prospective Study of Mortality," *Psychological Science* 14, no. 4 (2003), 320–327.

153 Morris A. Okun, Ellen Wan Yeung & Stephanie L. Brown, "Volunteering by Older Adults and Risk of Mortality: A Meta-Analysis," *Psychology and Aging* 28, no. 2 (2013), 564–577.

154 Michael J. Poulin, Stephanie L. Brown, Amanda J. Dillard & Dylan M. Smith, "Giving to Others and the Association Between Stress and Mortality," *American Journal of Public Health* 103, no. 9 (2013), 1649–1655.

155 Stephanie L. Brown, Dylan M. Smith, Richard Schulz, Mohammed U. Kabeto, Peter A. Ubel, Michael Poulin, et al., "Caregiving Behavior Is Associated with Decreased Mortality Risk," *Psychological Science* 20, no. 4 (2009), 488–494.

156 Elizabeth W. Dunn, Lara B. Aknin & Michael I. Norton, "Spending Money on Others Promotes Happiness," *Science* 319, no. 5870 (2008), 1687–1688.

157 Lara B. Aknin, Christopher P. Barrington-Leigh, Elizabeth W. Dunn, John F. Helliwell, Justine Burns, Robert Biswas-Diener, et al., "Prosocial Spending and Well-Being: Cross-Cultural Evidence for a Psychological Universal," *Journal of Personality and Social Psychology* 104, no. 4 (2013), 635–652.

158 Lara B. Aknin, Tanya Broesch, J. Kiley Hamlin & Julia W. Van de Vondervoort, "Prosocial Behavior Leads to Happiness in a Small-Scale Rural Society,"

Journal of Experimental Psychology: General 144, no. 4 (2015), 788–795.

159 Jorge Moll, Frank Krueger, Roland Zahn, Matteo Pardini, Ricardo de Oliveira-Souza & Jordan Grafman, "Human Fronto–Mesolimbic Networks Guide Decisions About Charitable Donation," *Proceedings of the National Academy of Sciences* 103, no. 42 (2006), 15623–15628.

160 Amy Wrzesniewski & Jane E. Dutton, "Crafting a Job: Revisioning Employees as Active Crafters of Their Work," *The Academy of Management Review* 26, no. 2 (2001), 179–201, 191.

161 Adam Grant, "8 Ways to Say No Without Ruining Your Reputation," Huffington Post, March 12, 2014. https://www.huffpost .com/entry/8-ways-to-say-no -without_b_4945289.

162 Netta Weinstein & Richard M. Ryan, "When Helping Helps: Autonomous Motivation for Prosocial Behavior and Its Influence on Well-Being for the Helper and Recipient," *Journal of Personality and Social Psychology* 98, no. 2 (2010): 222–244.

163 Martela et al., "Meaningfulness as Satisfaction of Autonomy, Competence, Relatedness, and Beneficence"; Martela & Riekki, "Autonomy, Competence, Relatedness, and Beneficence."

164 Richard Taylor, "The Meaning of Life," in *Life and Meaning: A Philosophical Reader*, ed. Oswald Hanfling, (Oxford: Blackwell, 1988), 39–48.

165 Lawrence C. Becker, "Good Lives: Prolegomena," *Social Philosophy*

and Policy 9, no. 2 (1992), 15–37, 20.

166 Letter from Tolstoy to N. N. Strakhov, April, 1876. Quoted in George Gibian, ed., *Anna Karenina—A Norton Critical Edition* (New York: W. W. Norton & Company), 751.

167 Rebecca J. Schlegel, Joshua A. Hicks, Laura A. King & Jamie Arndt, "Feeling Like You Know Who You Are: Perceived True Self-Knowledge and Meaning in Life," *Personality and Social Psychology Bulletin* 37, no. 6 (2011), 745–756.

168 See, e.g., Rebecca J. Schlegel, Joshua A. Hicks, Jamie Arndt & Laura A. King, "Thine Own Self: True Self-Concept Accessibility and Meaning in Life," *Journal of Personality and Social Psychology* 96, no. 2 (2009), 473–490.

169 See Ryan & Deci, *Self-Determination Theory*.

170 E.g., Valery Chirkov, Richard M. Ryan, R. M., Youngmee Kim & Ulas Kaplan, "Differentiating Autonomy from Individualism and Independence: A Self-Determination Theory Perspective on Internalization of Cultural Orientations and Well-Being," *Journal of Personality and Social Psychology* 84, no. 1 (2003), 97–110; Beiwen Chen, Maarten Vansteenkiste, Wim Beyers, Liesbet Boone, Edward L. Deci, Jolene Van der Kaap-Deeder, et al., "Basic Psychological Need Satisfaction, Need Frustration, and need Strength Across Four Cultures," *Motivation and Emotion* 39, no. 2 (2015), 216–236.

171 Chirkov et al., "Differentiating Autonomy from Individualism and Independence."

172 Richard M. Ryan & Edward L. Deci, "Self-Determination Theory and the Facilitation of Intrinsic Motivation, Social Development, and Well-Being," *American Psychologist* 55, no. 1 (2000), 68–78.

173 Robert Wuthnow, *Acts of Compassion: Caring for Others and Helping Ourselves* (Princeton, NJ: Princeton University Press, 1991).

174 Zeus states this to Aegistheus in Act II of Sartre's play *The Flies (Les Mouches)*, in Jean-Paul Sartre, *No Exit and Three Other Plays* (New York: Vintage Books, 1989), 102.

175 Frankl, *Man's Search for Meaning*, 104.

176 Ryan & Deci, *Self-Determination Theory*.

177 See chapter 10 in MacIntyre, *After Virtue*.

178 MacIntyre, *After Virtue*, 122.

179 Rawls called this the Aristotelian principle. John Rawls, *A Theory of Justice*, rev. ed. (Cambridge, MA: Harvard University Press, 2003).

180 People Are Awesome—YouTube channel boasts some five billion views, featuring videos where "ordinary people do extraordinary things."

181 Mihaly Csikszentmihalyi, *Flow: The Psychology of Optimal Experience* (New York: Harper Perennial, 1991).

182 Csikszentmihalyi, *Flow*, 4.

183 Pew Research Center, "Where Americans Find Meaning in Life."

184 It must be noted that as compared to the three other sources

discussed, competence has received the least direct research attention. So the experimental studies establishing its role as a key source of meaning in life are still something in need of doing. My argument for it being important for meaningfulness is thus mainly theoretical, although see Martela et al., "Meaningfulness as Satisfaction of Autonomy, Competence, Relatedness, and Beneficence."

185 Emily Esfahani Smith, *The Power of Meaning: Crafting a Life That Matters* (London: Rider, 2017), 229–230.

186 Tolstoy, *My Confession*, in *The Meaning of Life*, 14.

187 Seppälä, Emma 2016: *The Happiness Track: How to Apply the Science of Happiness to Accelerate Your Success*, (New York: Harper-Collins, 2016).

188 John Dewey, *How We Think* (New York: Cosimo, 2007), 219. Originally published in 1910.

189 Tolstoy, *My Confession*, in *The Meaning of Life*, 13.

190 Alan W. Watts, *The Tao of Philosophy*, edited transcripts (North Clarendon, VT: Tuttle Publishing, 2002), 77–78. Originally delivered as a lecture and transcribed by his son, Mark Watts.

191 Camus, *Myth of Sisyphus*, 63.

Baggini, Julian. "Revealed: The Meaning of Life," *The Guardian*, September 20, 2004. https://www.theguardian.com/theguardian/2004/sep/20/features11.g2.

Baumeister, Roy F., Kathleen Vohs, Jennifer Aaker, and Emily Garbinsky, "Some Key Differences Between a Happy Life and a Meaningful Life," *Journal of Positive Psychology* 8, no. 8 (2013).

Beauvoir, Simone de. *The Coming of Age*, trans. Patrick O'Brian. New York: W. W. Norton & Co., 1970/1996.

Burroughs, William S. *Naked Lunch*. New York: Grove Press, 1959/2001.

Camus, Albert. *Myth of Sisyphus*, trans. Justin O'Brien. New York: Vintage Books, 1955.

Carlyle, Thomas. *Sartor Resartus*. Oxford: Oxford University Press, 1834/1987.

——. *Heroes, Hero-worship, and The Heroic in History.* Chapman and Hall, London, 1840.

Coelho, Paulo. *Eleven Minutes*. New York: HarperCollins, 2005.

Covey, Stephen, A. Roger Merrill, and Rebecca R. Merrill. *First Things First*. New York: Fireside, 1995.

Emerson, Ralph Waldo. "Self-reliance," *The Essential Writings of Ralph Waldo Emerson*. New York: Modern Library, 1841/2000.

Haidt, Jonathan. *Happiness Hypothesis: Finding Modern Truth in Ancient Wisdom*. New York: Basic Books, 2006.

Hyman, Gavin. *A Short History of Atheism*. New York: I. B. Tauris & Co., 2010.

James, William. "Is Life Worth Living?" *The Will to Believe and Other Essays in Popular Philosophy*. New York: Dover Publications, 1897/1956.

Kierkegaard, Søren. "Repetition," *The Essential*

Kierkegaard, ed. Howard V. Hong & Edna H. Hong. Princeton, NJ: Princeton University Press, 1843/2013.

Krauss, Lawrence. "A Universe From Nothing," lecture delivered at AAI 2009 and available on YouTube. https://www.youtube.com/watch?v=7ImvlS8PLIo.

Mill, John Stuart. *Autobiography*. London: Penguin Books, 1873/1989.

Newton, Isaac. "General Scholium. An Appendix to the 2nd edition of the *The Mathematical Principles of Natural Philosophy*," 1713, http://www.newtonproject.ox.ac.uk/.

Nietzsche, Friedrich. *Nachgelassene Fragmente: Juli 1882 bis winter 1883–1884*, eds Giorgio Rolli and Mazzini Montinari. Walter de Gruyter, Berlin, 1977.

Nin, Anaïs. *The Diary of Anaïs Nin*, Vol. 2, 1934–1939, ed. Gunther Stuhlmann. Orlando, FL: Harcourt Brace & Company, 1967.

NPR, *All Things Considered*. "President Obama Is Familiar with Finland's Heavy Meal Scene. Are You?," May 17, 2016, https://www.npr.org/2016/05/17/478409307/president-obama-is-familiar-with-finlands-heavy-metal-scene-are-you.

Sagan, Carl. *Pale Blue Dot: A Vision of the Human Future in Space*. New York: Random House, 1994.

Seneca, "Letters, Book II, Letter XLVIII." Quoted in Stephen Salkever, ed., *The Cambridge Companion to Ancient Greek Political Thought*. New York: Cambridge University Press, c. 65 CE/2009.

Smith, Adam. *The Theory of Moral Sentiments*, ed. D. Raphael & A. Macfie. Indianapolis: Liberty Fund, 1759/1982.

Street, Sharon. "A Darwinian Dilemma for Realist Theories of Value," *Philosophical Studies* 127, no. 1, 2006.

Taylor, Charles. *A Secular Age*. Cambridge, MA: The Belknap Press of Harvard University Press, 2007.

Tolstoy, Leo. "Letter to Valerya," Quoted in Henri Troyat, *Tolstoy*, trans. Nancy Amphoux. New York: Grove Press, 1967.

Updike, John. "Picked-up Pieces, Moments from a Half Century of Updike," *The New Yorker*, February 1, 2009.

Vonnegut, Kurt. *Cat's Cradle*. New York: Dial Press, 1963/2010.

Worline, Monica C. and Jane E. Dutton. *Awakening Compassion at Work: The Quiet Power that Elevates People and Organizations*. Oakland, CA: Berrett-Koehler Publishers, 2017.

Wuthnow, Robert. *Acts of Compassion: Caring for Others and Helping Ourselves*. Princeton, NJ: Princeton University Press, 1991.

Aristotle, *Nicomachean Ethics,* trans. Robert C. Bartlett & Susan D. Collins. Chicago: University of Chicago Press, Chicago, 2012.

Beauvoir, Simone de. *Philosophical Writings,* ed. M. A. Simons. Urbana: University of Illinois Press, 2004.

Camus, Albert. *Myth of Sisyphus*, trans. Justin O'Brien. New York: Vintage Books, 1955.

Carlyle, Thomas. *Sartor Resartus.* Oxford: Oxford University Press, 1834/1987.

Ferry, Luc. *Learning to Live: A User's Manual.* Edinburg: Cannongate Books, 2010.

Frankl, Victor. *Man's Search for Meaning.* New York: Washington Square Press, 1963.

Haidt, Jonathan. *Happiness Hypothesis: Finding Modern Truth in Ancient Wisdom.* New York: Basic Books, 2006.

Hyman, Gavin. *A Short History of Atheism.* New York: I. B. Tauris & Co., 2010.

Kierkegaard, Søren. *Either/Or*, trans. Howard V. Hong & Edna H. Hong. Princeton, NJ: Princeton University Press, 1987.

Klemke, E. D. *The Meaning Life.* 4th ed. Edited by Steven M. Kahn. New York: Oxford University Press, 2017.

Landau, Iddo. *Finding Meaning in an Imperfect World.* New York: Oxford University Press, 2017.

MacIntyre, Alasdair. *After Virtue,* 3rd ed. Notre Dame, IN: University of Notre Dame Press, 2007.

May, Todd. *A Significant Life: Human Meaning in a Silent Universe.* Chicago: University of Chicago Press, 2015.

Metz, Thaddeus. *Meaning in Life: An Analytic Study.* Oxford: Oxford University Press, 2013.

Sartre, Jean-Paul. *Existentialism Is a Humanism,* trans. Carol Macomber. New Haven: Yale University Press, 2007.

RECOMMENDED READING

Smith, Emily Esfahani. *The Power of Meaning*. New York: Broadway Books, 2017.

Taylor, Charles. *The Ethics of Authenticity*. Cambridge, MA: Harvard University Press, 1991.

Taylor, Charles. *A Secular Age*, Cambridge, MA: The Belknap Press of Harvard University Press, 2007.

Tolstoy, Leo. *Confession*, trans. David Patterson. New York: W. W. Norton & Co., 1983.

Wilson, Colin. *The Outsider*. London: Pan Books Ltd., 1967.

FRANK MARTELA, PhD, is a philosopher and researcher of psychology specializing in the question of meaning in life. His articles have appeared in *Scientific American Mind* and *Salon,* and his work has been featured on Quartz and on the BBC. His research has been published extensively in numerous academic journals such as the *Journal of Personality, Journal of Positive Psychology, Journal of Happiness Studies, Metaphilosophy, Southern Journal of Philosophy, Motivation and Emotion,* and *Organization Studies.* He has spoken in front of more than one hundred audiences worldwide and has been invited to lecture at universities on four continents, including Stanford University and Harvard University. He's been interviewed by the *New York Times, Fitness, VICE,* and *Monocle.* Martela is based at Aalto University in Helsinki. Read more about him at www.frankmartela.com or follow him on Twitter @f_Martela.

ABOUT THE AUTHOR